Springer Tracts in Autonomous Systems

Volume 2

The Springer Tracts in Autonomous Systems (STRAUS) publish the state-of-the-art and advances in unmanned systems exhibiting attributes of autonomy, intelligence, autonomous functionality and resilience. The objective is to provide the foundations of the overall cycle from design and modeling of unmanned systems, to control and implementation exhibiting autonomy and self-organization amidst extreme disturbances, also functioning in uncertain and dynamic environments.

The STRAUS series is the first of its kind in that it provides 'metrics' to evaluate autonomy and autonomous functionality and resilience, directly applicable to real complex engineering systems.

Books to be published are suitable for the scientist and engineer, practitioner and end-user. They are also suitable for graduate students and researchers who wish to conduct research in the topical areas covered by the series. Within the scope of the series are monographs, lecture notes, selected contributions from specialized conferences and workshops, as well as selected PhD theses.

More information about this series at http://www.springer.com/series/16447

Yufeng Yue · Danwei Wang

Collaborative Perception, Localization and Mapping for Autonomous Systems

Springer

Yufeng Yue
School of Automation
Beijing Institute of Technology
Beijing, China

Danwei Wang
School of Electrical
and Electronic Engineering
Nanyang Technological University
Singapore, Singapore

ISSN 2662-5539 ISSN 2662-5547 (electronic)
Springer Tracts in Autonomous Systems
ISBN 978-981-15-8859-4 ISBN 978-981-15-8860-0 (eBook)
https://doi.org/10.1007/978-981-15-8860-0

This Springer imprint is published by the registered company Springer Nature Singapore Pte Ltd.
The registered company address is: 152 Beach Road, #21-01/04 Gateway East, Singapore 189721, Singapore

To my wife, Siqi Zhang
Yufeng Yue

To my wife, Ming Luo, and Our children:
Anmin, Andai and Anyu
Danwei Wang

Preface

For hundreds of years, creating machines that can perform tasks autonomously have been the pursuit of many pioneers. Thanks to the development of latest technology, autonomous systems and robots have been gradually applied in various scenarios such as industry, daily life, and transportation. As robots become more common, they are expected to interact with each other, share information, and execute tasks collaboratively. This idea is very similar to human behavior in many situations. For example, imaging a team is playing basketball, members are assigned with tasks such as defensing, scoring, rebounding, etc. Each team member will complete different tasks and act as a collaboration system. The same idea applies to a system that is composed of multiple autonomous robots. Collaborative autonomous systems have benefits of multi-perspective perception, high efficiency, and robustness to single agent failure. In the past few decades, extensive research has been conducted on the development of collaborative autonomous systems, involving a wide range of fields such as perception, localization, planning, navigation and control.

A majority of the published literature on collaborative autonomous systems are under well-defined and constrained environments. Therefore, current collaborative robot systems are not fully capable of autonomously operating in complex environments. Due to limited sensing capabilities, each robot only has partial information about the surroundings. In such cases, the fundamental challenge is the deployment of a collaborative system which can perform distributed perception, localization and mapping, deal with the incomplete information and uncertainty. More specifically, robots need to analyze, understand and describe the environment with a proper form. Then, all robots have to share local maps to estimate the accurate relative position between robots and generate a consistent global map. The development of collaborative perception systems has a wide range of application fields. For a group of autonomous vehicles, sharing local maps among them will avoid accidents caused by occlusion, and update the global map of the city more efficiently.

What is the subject and benefit? This book presents the breakthrough and cutting-edge progress for collaborative perception, localization and mapping. The book proposes a novel framework of "multimodal perception-relative localization collaborative mapping" for collaborative robot systems. The organization of the book allows the readers to analyze, model, and design collaborative perception technology for autonomous robots. First of all, the point set registration algorithm is proposed for 3D map matching in a structured environment. Then, a novel hierarchical probabilistic map fusion framework is studied, where low-level points and high-level geometry features are combined. When the autonomous systems are equipped with heterogeneous sensors, a general probabilistic framework is proposed, which is independent of sensor types. In a challenging all-weather and dynamic environment, day and night collaborative dynamic mapping is proposed based on multimodal sensors. Finally, with the rapid development of convolutional neural networks, this book promotes collaborative perception to a higher semantic level and proposes a collaborative semantic mapping algorithm.

Thus, the purpose of this book is to present the basic foundation in the field of collaborative robot systems. The book provides fundamental theory and technical guidelines for collaborative perception, localization and mapping. In addition, the book will significantly promote the development of autonomous systems from individual intelligence to collaborative intelligence. In order to improve the readability, extensive simulations and real experiments results are presented along different chapters.

Who is this book for? For engineers, graduate students and researchers in the fields of autonomous systems, robotics, computer vision and collaborative perception, this book can arouse their great interest. Nevertheless, experts in the fields of mathematics, electronic engineering, automation, or artificial intelligence can also find new ideas in the book. In order to fully understand the concepts presented in this book, readers should have a basic understanding of probabilistic theory and robotics.

How this book is best used? The book provides theoretical basis and technical implementation guidelines for collaborative perception, localization and mapping, which can be implemented by the reader and can be a start point for future research. As a research monograph, the book is aimed to provide the fundamental concept and method of analyzing and modelling collaborative systems, which can be a reference book for researchers from different fields. As a practical guide, the book provides technical implementation guidelines, pseudocode and flow diagrams for researchers to develop their own systems.

Beijing, China Yufeng Yue
Singapore, Singapore Danwei Wang

Contents

Chapter 1
Introduction

1.1 Background

1.1.1 Motivations

In the field of robotics and artificial intelligence, autonomous robot has been a highly active research area, which enables a robot to perform behaviors with a high degree of autonomy. With the continuing development in the field of mobile robotics, in particular, using multiple robots to autonomously explore and map an environment has caught the attention of many robotics researchers and practitioners (see Fig. 1.1). Utilizing a group of robots have found significant advantages compared to their single robot counterpart: robustness to single robot failure, quicker exploration of environments in time critical search and rescue missions, execution of tasks of high complexity, reduction of human risks, and costs associated to disaster response. Accurate localization and mapping is a crucial competency for enabling collaboration between multiple mobile robots.

In the past few decades, extensive research has been conducted on the development of collaborative autonomous systems, several review papers have summarized the latest results in this area [1–3]. Employing multiple robots introduces a new set of challenges such as multi-robot perception [4, 5], multi-robot simultaneous localization and mapping (SLAM) [6–8], distributed system [9–11] and coordinated exploration [12–14], to name a few. Among the above mentioned research hot spots, collaborative perception, localization and mapping [15, 16], which is to combine and distribute the data perceived by individual robots to estimate the relative position and reconstruct a comprehensive understanding of the environment, is one of the most critical topics. As various intelligent robots are entering human daily life to perform different tasks like package delivery, security patrolling and taking care of human. In these restricted and dynamic environments, such as dense urban area and large communities, prior maps may not be available, and Global Positioning System

Y. Yue and D. Wang, *Collaborative Perception, Localization and Mapping for Autonomous Systems*, Springer Tracts in Autonomous Systems 2,
https://doi.org/10.1007/978-981-15-8860-0_1

Fig. 1.1 Teams of autonomous robots designed to perform reconnaissance missions

(GPS) signals may be disrupted. Therefore, multiple robots must perform intelligent perception and environment understanding, share and match their individual observations for relative localization and mapping. This book focuses on the collaborative perception, localization and mapping problem for multiple autonomous robots.

1.1.2 Challenges

Many challenges exist in collaborative systems and attempts have been made from different aspects to address them, such as dynamic environment understanding [17, 18] estimating the relative transformation between the maps [19], accounting for the uncertainty of the relative transformation [4], integrating the information from neighboring robots [20], dealing with complexity issues[10], manage the communication [15], updating maps in the global perspective [21], and many more. This book is motivated by the fact that there are several challenges in addressing some of the related issues of collaborative perception, localization and mapping. The major challenge in these systems is to design a general framework for collaborative systems, which covers multimodal perception, relative localization, collaborative mapping (see Fig. 1.2):

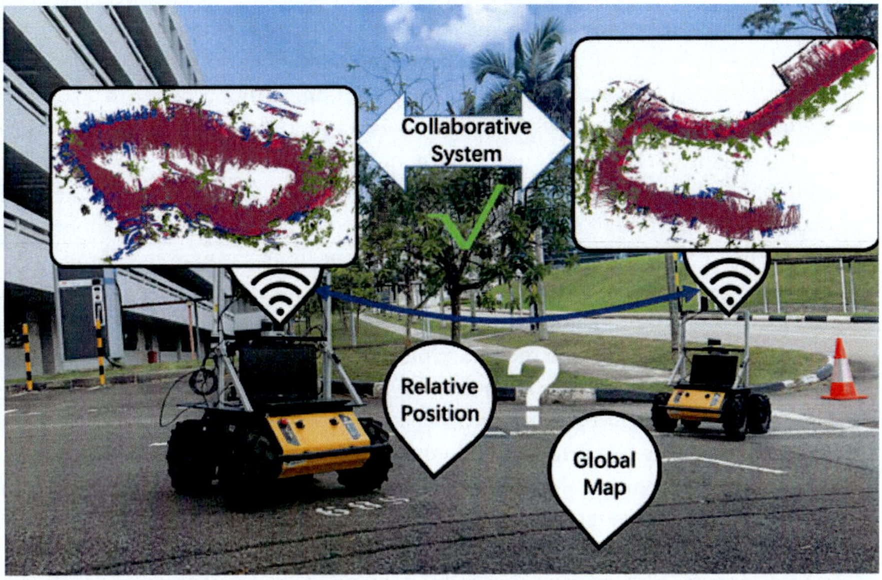

Fig. 1.2 Demonstration of semantic map matching with the collaborative autonomous systems, which integrates three sub-modules of "multimodal perception-relative localization-collaborative mapping"

1. *General framework for collaborative system*. The research of exploiting multimodal perception, relative localization, collaborative mapping in both single robot and multi-robot levels has not been studied extensively. In general, both levels are composed of estimating the relative transformation and integrating the partial map information, which are usually tackled separately. In summary, previous approaches have not provided comprehensive analysis of the integrated map fusion system on map uncertainty modeling, probabilistic map matching, transformation evaluation, accurate map merging and communication management.
2. *Probabilistic map matching*. Most of the existing map matching algorithms belong to the scope of iterative dense registration, which is designed for raw sensor data registration on point/voxel-wise level. Due to the intrinsic nature difference between raw sensor data and 3D occupancy grid maps, the direct use of dense registration algorithms in 3D occupancy grid map matching results in sub-optimal voxel associations thus sub-optimal matching. To increase the accuracy and convergence, available information like geometry information, and occupancy probability should be incorporated in the matching process.
3. *Transformation evaluation and inconsistency detection*. The majority of the research is focused on generating accurate maps and post-processed optimization. However, the generated map does not guarantee to be foolproof and may generate inconsistent maps. Fusing erroneous local maps results in an inconsistent global map, which renders the accurate execution of autonomous tasks infeasible. Therefore detecting inconsistency in both local maps and fused global maps is

vital for multi-robot missions. The capability to detect inconsistency in the generated maps allows robots either to perform an efficient optimization algorithm, or to return to a safe location for recovery without getting trapped.
4. *Collaborative mapping using heterogeneous sensors*. For a group of robots in arbitrary (structured/semi-structured/unstructured) environments, the robot might be equipped with multiple sensors. To utilize their complementary characteristics, map generated with heterogeneous sensors need to be fused. Existing approaches have relied on point set registration based algorithms for map matching, where hard one-to-one correspondence relationship is assigned. Since the intrinsic characteristics (density/range/field of view) of the map produced by heterogeneous sensors are very different, a more flexible and general map registration algorithm is required and should be independent of sensor types.
5. *Semantic Understanding in all-weather environments*. Performing collaborative semantic mapping is a critical challenge for cooperative robots to enhance their comprehensive contextual understanding of the surroundings. Current 3D geometry mapping approaches only contain geometry information, which limit the application of robots in high level tasks. In all-weather dynamic environment, robots must be able to recognize dynamic objects and collaboratively build a global map. Therefore, to enhance the perception capability of a group robots, individual robots have to perform semantic understanding and generate a global representation composed of geometry information and semantic context.

1.2 Objective of This Book

The main objective of this book is proposing the collaborative robots perception, localization and mapping framework so as to improve the accuracy, efficiency and robustness. The framework will act as the theoretical basis for multi-robot coordination missions, which enables the team of robots to work with heterogeneous sensors in arbitrary (structured/semi-structured/unstructured) environments. The proposed solutions need to overcome key practical issues such as uncertainty modeling, probabilistic map matching, transformation evaluation, information integration, semantic understanding under limited communication bandwidth and computational power.

As the book is concerned about the objectives mentioned above, a novel framework of "multimodal perception-relative localization-collaborative mapping" for collaborative robot systems is proposed. First of all, the point set registration algorithm is proposed for 3D map matching in structured environment. Then, a novel hierarchical probabilistic map fusion framework is studied, where low level points and high level geometry features are combined. When the autonomous systems are equipped with heterogeneous sensors, a general probabilistic framework is proposed, which is independent of sensor types. In a challenging all-weather and dynamic environment, day and night collaborative dynamic mapping is proposed based on multimodal sensors. Finally, with the rapid development of convolutional neural networks, this book promotes collaborative perception to a higher semantic level and proposes a collaborative semantic mapping algorithm.

1.3 Preview of Chapters

The remainder of this book is organized as follows.

- Chapter 2 gives an extensive survey of collaborative robot systems. The main approaches published in the last ten yeas with regard to our common thread are reviewed.
- Chapter 3 details the probabilistic point set registration approach for fusion of 3D occupancy grid maps. This chapter first presents Occupancy Iterative Closet Point algorithm to register the maps. Then, the evaluation of the computed relative transformation and the fusion of probabilities is presented. The accuracy of the proposed algorithm is evaluated using maps generated from both simulated and real environments.
- Chapter 4 explains hierarchical probabilistic fusion framework for matching and merging of 3D occupancy maps. The uncertainty of 3D occupancy map data at voxel level and structural level are analyzed in depth. The two-level probabilistic map matching framework based on the uncertainty modeling is formulated. Then, the statistical testing for transformation estimation and the relative entropy-based map merging approach are proposed. The results verify the accuracy and efficiency of the proposed 3D occupancy map fusion approach, and exhibit the improved convergence in these scenarios.
- Chapter 5 introduces a general probabilistic framework for collaborative 3D mapping using heterogeneous sensors. The chapter formulates the joint probability of the global map and the set of relative transformations between the local maps. Then, it also proposes a time-sequential map merging framework to share map information among robots. The proposed approach is evaluated in real-time experiments in various environments with heterogeneous sensors, which shows its robustness and generality in 3D map fusion for multi-robot mapping missions.
- Chapter 6 addresses dynamic collaborative mapping based on multimodal environmental perception. For each mission, robots first apply heterogeneous sensor fusion model to detect humans and separate them to acquire static observations. Then, the collaborative mapping is performed to estimate the relative position between robots and local 3D maps are integrated into a globally consistent 3D map. The experiment is conducted in the day and night rainforest with moving people. The results show the accuracy, robustness, and versatility in 3D map fusion missions.
- Chapter 7 proposes a hierarchical collaborative probabilistic semantic mapping framework, where the problem is formulated in a distributed setting. At single robot level, the semantic point cloud is obtained by a fusion model combining information from heterogeneous sensors and is used to generate local semantic

maps. At collaborative robots level, local maps are shared among robots for global semantic map fusion. The experimental results on the UAV (Unmanned Aerial Vehicle) and UGV (Unmanned Ground Vehicle) platforms show the high quality global semantic map, demonstrating the accuracy and utility in practical missions.

- Chapter 8 summaries this book. The contributions of this work together with the expected future work are provided.

References

1. Saeedi S, Trentini M, Seto M, Li H (2016) Multiple-robot simultaneous localization and mapping: a review. J Field Robot 33(1):3–46
2. Darmanin RN, Bugeja MK (2017) A review on multi-robot systems categorised by application domain. In: 2017 25th mediterranean conference on control and automation (MED), pp 701–706. https://doi.org/10.1109/MED.2017.7984200
3. Gautam A, Mohan S (2012) A review of research in multi-robot systems. In: 2012 IEEE 7th international conference on industrial and information systems (ICIIS), pp 1–5. https://doi.org/10.1109/ICIInfS.2012.6304778
4. Dong J, Nelson E, Indelman V, Michael N, Dellaert F (2015) Distributed real-time cooperative localization and mapping using an uncertainty-aware expectation maximization approach. In: 2015 IEEE international conference on robotics and automation (ICRA), pp 5807–5814
5. Yue Y, Yang C, Zhang J, Wen M, Wu Z, Zhang H, Wang D (2020) Day and night collaborative dynamic mapping in unstructured environment based on multimodal sensors. In: 2020 IEEE international conference on robotics and automation (ICRA)
6. Rashid AT, Frasca M, Ali AA (2015) Multi-robot localization and orientation estimation using robotic cluster matching algorithm. Robot Auton Syst 63:108–121
7. Hussein M, Renner M, Watanabe M, Iagnemma K (2013) Matching of ground-based lidar and aerial image data for mobile robot localization in densely forested environments. In: 2013 IEEE/RSJ international conference on intelligent robots and systems (IROS). IEEE, pp 1432–1437
8. Yue Y, Zhao C, Li R, Yang C, Zhang J, Wen M, Wang Y, Wang D (2020) A hierarchical framework for collaborative probabilistic semantic mapping. In: 2020 IEEE international conference on robotics and automation (ICRA)
9. Stroupe AW, Martin MC, Balch T (2001) Distributed sensor fusion for object position estimation by multi-robot systems. In: Proceedings 2001 ICRA. IEEE international conference on robotics and automation (Cat. No.01CH37164), vol 2, pp 1092–1098
10. Cunningham A, Indelman V, Dellaert F (2013) Ddf-sam 2.0: consistent distributed smoothing and mapping. In: 2013 IEEE international conference on robotics and automation, pp 5220–5227. https://doi.org/10.1109/ICRA.2013.6631323
11. Wang Y, Shan M, Yue Y, Wang D (2019) Vision-based flexible leader-follower formation tracking of multiple nonholonomic mobile robots in unknown obstacle environments. IEEE Trans Control Syst Technol
12. Trautman P, Ma J, Murray RM, Krause A (2015) Robot navigation in dense human crowds: statistical models and experimental studies of human robot cooperation. Int J Robot Res 34(3):335–356
13. Gregoire J, Čáp M, Frazzoli E (2017) Locally-optimal multi-robot navigation under delaying disturbances using homotopy constraints. Auton Robots
14. Burgard W, Moors M, Stachniss C, Schneider FE (2005) Coordinated multi-robot exploration. IEEE Trans Robot 21(3):376–386

15. Niedfeldt PC, Speranzon A, Surana A (2015) Distributed map fusion with sporadic updates for large domains. In: 2015 IEEE international conference on robotics and automation, pp 2806–2813
16. Bonanni TM, Corte BD, Grisetti G (2017) 3-D map merging on pose graphs. IEEE Robot Autom Lett 2(2):1031–1038. https://doi.org/10.1109/LRA.2017.2655139
17. Yang C, Wang D, Zeng Y, Yue Y, Siritanawan P (2019) Knowledge-based multimodal information fusion for role recognition and situation assessment by using mobile robot. Inf Fusion 50:126–138
18. Yang C, Yue Y, Zhang J, Wen M, Wang D (2018) Probabilistic reasoning for unique role recognition based on the fusion of semantic-interaction and spatio-temporal features. IEEE Trans Multimed 21(5):1195–1208
19. Yue Y, Yang C, Wang Y, Senarathne PCN, Zhang J, Wen M, Wang D (2019) A multilevel fusion system for multirobot 3-d mapping using heterogeneous sensors. IEEE Syst J 14(1):1341–1352
20. Saeedi S, Paull L, Trentini M, Seto M (2014) Group mapping: a topological approach to map merging for multiple robots. Robot Autom Mag IEEE 21(2):60–72
21. Yue Y, Senarathne PN, Yang C, Zhang J, Wen M, Wang D (2018) Hierarchical probabilistic fusion framework for matching and merging of 3-d occupancy maps. IEEE Sens J 18(21):8933–8949

Chapter 2
Technical Background

In this chapter, an extensive survey related to existing approaches of collaborative perception, localization and mapping is provided. Among these solutions, estimating relative transformation, integrating the information from neighboring robots, dealing with complexity issues, managing the communication network, and updating maps in the global perspective are the majority. The main approaches published in the last ten yeas with regard to multi-robot mapping method are reviewed.

The remaining part of the chapter proceeds as below. In Sect. 2.1, single-robot SLAM and multi-robot SLAM problems are reviewed and main approaches to collaborative mapping are summarized. Section 2.2 summarizes the state of the art of the map matching solutions, where registration of raw sensor, homogeneous map matching and heterogeneous map matching are thoroughly reviewed. Sect. 2.3 describes the distributed map sharing, map inconsistency detection and probabilistic information fusion, which are categorized into the scope of the map merging problem.

2.1 Collaborative Perception and SLAM

2.1.1 Single Robot Perception and SLAM

This section presents various approaches existed to solve single robot semantic perception and SLAM problem. The single robot SLAM based on geometry features is summarized first, and then moves to perception and SLAM with semantic information. The reason for this section is to summarize the available solutions and to have a better understanding of multi-robot coordination problem.

Y. Yue and D. Wang, *Collaborative Perception, Localization and Mapping for Autonomous Systems*, Springer Tracts in Autonomous Systems 2,
https://doi.org/10.1007/978-981-15-8860-0_2

SLAM describes a problem of placing a robot in an unknown environment without GPS information, which has to map the environment and localize itself with respect to the map. Both the trajectories of the robot and the map should be estimated online by using the external sensors like cameras, lasers, radars, or internal sensors like IMUs, wheel encoders. Solving the SLAM problem and providing a reliable online algorithm is a key technology for autonomous robots.

Assuming that robot poses are $x_{1:k} = \{x_1, x_2, \ldots, x_k\}$, where x_k is the current post of the robot at time k. The map at time k is denoted by the set $m_k = \{m_{k,1}, m_{k,2}, \ldots, m_{k,l_k}\}$, and l_k represents the number of voxels(landmarks) at time k. The observations made by the robot are $z_{1:k} = \{z_1, z_2, \ldots, z_k\}$, and $z_k = \{z_{k,1}, z_{k,2}, \ldots, z_{k,n_k}\}$ is the sensor observation set perceived at time k, while n_k presents the number of observations. The control sequences are $u_{1:k} = \{u_1, u_2, \ldots, u_k\}$, and u_k is the control command applied to the robot at time k.

An online SLAM problem is defined to only estimate the posterior over the current position x_k and map m_k, which is:

$$p(x_k, m_k | z_{1:k}, u_{1:k}, x_0) \tag{2.1}$$

Based on some initial job in [1], extended kalman filter (EKF) is proposed in [2–4]. EKF models the feature map as vectors and linearize the nonlinear function to achieve efficiency. However, solving the state vector will be highly computational expensive and the linearization also cause large errors if the environment is highly non-linearized. Then Fast SLAM 1.0 and 2.0 are proposed in [5] and [6], respectively, which applies Monte Carlo Sampling with particles to recursively estimate the non-linear and non-Gaussian environment. However, Fast SLAM has to maintain a numerous number of particles and is prone to impoverishment by resampling. In addition to EKF and Fast SLAM, a notable approach called sparse extended information filters (SEIF) is proposed in [7], which can update equations and execute in constant time irrespective of the size of the map. For all the SLAM methods mentioned above, they all have to deal with the data association problem, which means to determine the correspondence between measurements and landmarks. The data association problem is always solved with a measurement model [8], scan matching [9] or an image feature extraction algorithm [10]. To incorporate the data association problem to the Bayesian filter, a random finite set(RFS) based on probability hypothesis density(PHD) filter is proposed in [11].

Full SLAM problem involves a lot of computational power, and may not process online. Then, a full SLAM problem is to estimate the trajectories of the robot at each time step $x_{1:t}$ and the posterior over the map m, which is:

$$p(x_{1:k}, m_k | z_{1:k}, u_{1:k}, x_0) \tag{2.2}$$

A full SLAM algorithm has to optimize the overall posterior of robot trajectories and map, and minimize the error caused by measurements and control input. In [12], a full SLAM algorithm called graph SLAM is proposed, which optimize the overall trajectories and landmarks by nonlinear minimization offline. The graph SLAM transfers the posterior into a graph network, and then reduce the variables to get a sparse matrix for minimization solving. In [13], an incremental smoothing and mapping approach is proposed based on fast incremental matrix factorization to perform full SLAM online. As shown in [14], graph SLAM is extended to large scale using aerial images as prior information. To achieve global convergence, loop closure detection is one of the key challenge. Loop closure means when a robot revisits a place, the robot is able to recognize. In large environment, a new edge will be added to the graph when a loop closure is detected, which will largely reduce the error caused by autonomous exploration. In [15], a switchable constraint is proposed for robust loop closure detections against outliers like data association errors and false positive. Latif et al. [16] extends loop closure to incremental searching and multi-sessions with the applying of g2o [17], which is a famous graph slam package.

Visual SLAM has a long history. Early visual SLAM systems are originated from conventional SLAM, where sequential filtering is used [18]. Given robot control signals and feature measurements, by assuming that the state-space estimation and measurements follow Gaussian distribution, a probability framework is formulated to deal with errors from different sources (e.g. motion model and observation model). Current visual SLAM takes SLAM as graph optimization process based on bundle adjustment and loop closing technique. Double window and multi window optimization which are well-suited for different environments are also explored [19]. Instead of taking all the frames and all the features into account, the concept of keyframe is commonly used to maintain the sparsity of graph. A frame is selected as keyframe only if the overlap between current frame and previous keyframe is lower than a threshold [20]. The connections between all the nodes of the graph are decreased dramatically compared with filtering approach. Real time implementation is no longer too difficult [21]. A very detailed comparison study on filtering and optimization approach has been given in [22].

In recent years, Convolutional neural networks (CNN) has become mainstream in high-dimensional applications, achieving satisfying results in a series of computer vision tasks such as image classification, segmentation, and object detection [23]. The ability to interpret the scene is an important ability for a robot to interact with the environment. Semantic segmentation labels each pixel of an image with a class label and thus provides a detailed semantic annotation of the surroundings. For semantic segmentation tasks, a novel fully connected networks (FCN) architecture Segnet is presented [24], where the decoder up-samples its lower resolution input as feature map. Refinenet [25] is proposed to exploit all the information for high-resolution image prediction. In Deeplab [26], multiple parallel atrous convolutions with different rates are involved to process the feature map. In this chapter, Deeplab is employed to process the image data perceived by mobile robots.

To realize single robot semantic mapping, each robot needs to localize itself accurately in the environment. Up to now, single robot SLAM (Simultaneous Localization and Mapping) is considered as a well-studied problem. Current approaches mainly consists of visual SLAM [27], and graph SLAM [28]. However, traditional SLAM only restores the geometry information of the surroundings.

With the rapid development of deep learning, semantic mapping has attracted a lot of attention. A comprehensive review for single robot semantic mapping is summarized in [29]. A hierarchical framework is proposed to combine spatial information and geometric maps in [30], but is only used in a 2D environment. As the availability of high-quality 3D sensors increases, a framework of generating volumetric 3D environment model is proposed in [31]. Then, an incremental semantic 3D mapping system for large-scale environments using scrolling occupancy grid is proposed in [32]. To improve the efficiency, octree based multi-label 3D semantic mapping algorithm is proposed in [33]. Since semantic information is also important for navigation [34], an integrated framework is presented that aims to use semantic map to provide a hierarchical navigation solution for the first time [35]. More recently, a semantic SLAM system [36] is presented that uses object-level entities to construct semantic objects and integrate them into the semantic SLAM framework. To perform mapping in dynamic environments, a stereo-based dense mapping algorithm is proposed [37]. Regarding the reconstruction of moving objects, [38] incrementally fuses sensor observations into a consistent semantic map. Aforementioned approaches promote the development of single robot semantic mapping. However, single robot mapping mainly fuses incoming raw sensor data, while collaborative mapping takes local maps as input. This inherent difference prevents existing algorithms from being directly applied to collaborative semantic mapping, and therefore requires novel frameworks and theoretical formulas.

2.1.2 Multi-Robot SLAM

Since single robot SLAM is difficult enough, enabling a group of robots to perform tasks autonomously brings challenges at a higher level. Coordination among multiple robots enable them to perform difficult tasks in challenging environments more efficiently and reliably, some of the applications include: earthquake surveillance [39], forest mapping [40], detection and tracking of moving objects [41], intelligent transportation [42] and underwater mapping [43]. The approaches reviewed in last section is the base for multi-robot SLAM, and additional challenges include robot-wise communication, relative position estimation, loop closure detection, uncertainty updating and data fusion, which is summarized in Table 2.1.

Assuming that robot poses are $x_{1:k}^{(r)} = \{x_1^{(r)}, x_2^{(r)}, \ldots, x_k^{(r)}\}$, and $x_k^{(r)}$ represents the pose of the robot r with in its local coordinate frame at time k. The map at time k is given as $m_k^{(r)} = \{m_{k,1}^{(r)}, m_{k,2}^{(r)}, \ldots, m_{k,l_k^{(r)}}^{(r)}\}$, where $l_k^{(r)}$ represents the number of voxels (landmarks) at time k. Then the global map M at time k is given as $M_k = \{m_k^{(1)}, m_k^{(2)}, \ldots, m_k^{(r)}\}$. The sensor observations made by each robot are given as $z_{1:k}^{(r)} = \{z_1^{(r)}, z_2^{(r)}, \ldots, z_k^{(r)}\}$, and $z_k^{(r)} = \{z_{k,1}^{(r)}, z_{k,2}^{(r)}, \ldots, z_{k,n_k^{(r)}}^{(r)}\}$ denotes the sensor observation set perceived from robot r at time k, while $n_k^{(r)}$ is the number of observations at this time stamp. The control inputs given to the robots are $u_{1:k}^{(r)} = \{u_1^{(r)}, u_2^{(r)}, \ldots, u_k^{(r)}\}$, and $u_k^{(r)}$ represents the control command given to robot

Table 2.1 The challenges for multiple robots

Type	Descriptions
Communications	The way of robots communicating with each other, it can be centralized and distributed
Relative position	When multiple robots are performing tasks, the robot need to know the position of other robots in a global coordinate frame
Complexity	The complexity of operating multiple robots, it is important to enable the robots to perform in real time with computational power limitation
Data fusion	One of the most important reason of applying multiple robots is to fuse the measurements generated by individual robots and provide the global and full map of the explored environment

r at time k. Then, a full muli-robot SLAM problem is to estimate the continuous trajectories of each robot r at time step $x_{1:k}^{(1,\cdots,r)}$ and the update global map M_k, which is:

$$p(x_{1:k}^{(1,\cdots,r)}, M_k | z_{1:k}^{(1,\cdots,r)}, u_{1:k}^{(1,\cdots,r)}, x_0^{(1,\cdots,r)}) \tag{2.3}$$

To extend single robot SLAM to multi-robot SLAM, various solutions based on EKF have been proposed, such as merge sensor and navigation information from multiple robots in [42, 44, 45]. Extension from a single robot to multiple robots using EKF is straightforward, however, it suffers from linearization error and exponentially growing computational burden. Multi-robot based on particle filters is proposed in [46], which starts with not known prior information and integrates all data into a single common map. The particle filters based approach is further improved to deal with limited communication in large scale environment [47]. Alternatively, a graph SLAM based multi-robot coordination algorithm is proposed in [48], which is based on stochastic estimation and central convergence is proved. In [49], the author utilizes a cost function to enable a team of robots to explore an unknown environment in a minimum cost of reaching a target point. In [50], a neural network-based multi-robot SLAM is proposed. To handle dynamic environment with multiple robots, multi-robot coordination based on random finite set is proposed [51].

Aforementioned approaches mainly work in centralized manner, distributed multi-robot coordinate brings the challenges of data sharing, relative localization, loop closing and communication management. In [52], the author presents a distributed data fusion network for robots to share map information with neighbors under the limited communication bandwidth. In [53], distributed real-time cooperative localization and mapping is achieved using an uncertainty-aware expectation maximization approach. In [54], the paper addresses the problem of determining the relative pose between robots. Specifically, the author proves the minimum number of distance measurements required for determining all six possible solutions for 2D robot SLAM is three. In [55], the paper shows the loop closure detection with multiple robots. In [56], the author presents a map fusion algorithm of multi-robot slam using reinforcement learning. Submap based SLAM has been studied [57, 58], which combines with loop closure detection [59].

Table 2.2 The types of data fusion

Data Type	Descriptions
Raw sensor data	fuse whatever the sensor observed, like point cloud, images, odometry
Volumetric maps	fuse the maps generated by the robots using the sensor measurement, which can be 2D map and 3D map
Topology Stucture	fuse the topology structure extracted from the maps, it can be triangulation, voronois

2.1.3 Multi-Robot Map Fusion

Detailed 3D mapping of unknown environments is a major research problem in mobile robotics that has garnered significant attention in recent years with the improvements in 3D sensors, 3D mapping algorithms, and processing systems [60]. Utilizing multiple robots to conduct the mapping is going to be inevitable. Many modern autonomous mobile robotic applications ranging from environmental mapping [60, 61], surveillance [39] missions to autonomous vehicles require a detailed 3D map of the environment as a prerequisite for their optimal operation. Since the operating environment gets larger, efficiency and robustness of map generation become a critical factor [41, 62].

The major challenge in these systems is to fuse the maps generated by individual robots (i.e., local maps) into a globally consistent map (i.e., a global map). This is further exacerbated when the robots are required to operate in a networked environment with real or semi-real time mapping requirement, where the local maps are to be transferred over the network for map fusion. Equipped with various sensors, the laser sensor data and camera image data are very different at the raw sensor data level [63], where accurate transformation between the maps need to be calculated. Under the condition of limited bandwidth and coverage of communication channels [52], large amount of data is hard to transfer in real time [64]. Real or semi-real time global mapping with limited network bandwidth requires a careful consideration for the choice of map type that balances the requirements of detailed 3D mapping, support for transfer over limited communication and their utility for robust map fusion [65]. Some researches have been done to address data fusion among multiple robots which can be summarized into three different types (see Table 2.2). Raw sensor data [66, 67], volumetric maps [68–70] and topological maps [71, 72] are some of the different types of maps that can be used to represent the 3D environment.

While raw sensor data provides more detailed information about the environment, they are extremely challenging to store, manage and process and even more so to transfer over networks with limited connectivity. Therefore, the point cloud is not an ideal choice of map type for real-time multi-robot mapping missions. However, the rich details in these maps have encouraged the development of 3D map fusion algorithms [66], albeit the off-line high computational effort. Topological maps, on the other hand, are easier to process and transfer over the network due to their lower memory footprint. The literature contains map fusion algorithms involving 2D topological maps [65, 71, 72]. However, the high level of abstraction in these maps do not lend well for detailed 3D environment mapping and hence for 3D map fusion. Metric grid maps such as volumetric maps [31, 73] can represent the 3D environments in sufficient detail without sacrificing the memory or processor usage. Combined that with the availability of compression schemes that allow real-time transferring over limited bandwidth makes them the ideal choice of map type for multi-robot mapping missions. Octomap [31] is a volumetric map type with a high compression ratio that can represent any type of 3D environment, hence is selected as the choice of map type in this research.

2.2 Data Registration and Matching

The objective of map matching is to estimate the relative transformation between the robots. The two robots build local maps with coordinate frame L_a and L_b, then the local maps should be transformed into global coordinate. Over the last years, various research groups have developed efficient solutions to the map matching problem. Map matching can be performed at different levels, and should consider the trade-off between efficiency and quality. These techniques range from 2D grid map matching [71, 74, 75], to sparse feature descriptors [76, 77], to full 3D dense occupancy map registration [78, 79]. Here, the work focuses on map matching and makes the assumption of the local map generated by each robot is consistent. As the majority of the initial solutions to 3D map fusion is inspired by registration algorithms, sensor data based dense registration algorithms are reviewed first. Then, the approaches of map matching are reviewed, where the strengths and limitations are carefully discussed.

2.2.1 Registration of Sensor Data

To establish the point-wise correspondence, dense registration algorithms are directly operated on the data obtained by range sensors like Kinect, laser scanners. The objective is to find a 3D rigid transformation to register the consecutive sensor scans. The well known Iterative Closet Point (ICP) [80] is the most significant algorithm, which iteratively searches for corresponding pairs and computes the transform that minimizes the Euclidean distance. Since the introduction of ICP, many variants based on ICP have been proposed to improve the aspects like pre-filtering [81, 82], data association [83–86], outlier removal [87, 88] and error function optimization [89–92].

Prior to registration, scan data is usually preprocessed to extract features like planar patches [81], or down-sampling [82] to increase the efficiency. To find the data association: ICP assumes the points in two sources have physical spatial correspondence, however, this is violated in the presence of sensor noise and limited sensor density. To model the sensor noise, a probabilistic version is proposed in [93], where points are represented by a normal distribution. In [84], probabilistic data association is applied to align two point clouds with different densities. To deal with the sensor data with low sampling resolution, [94] represents the point cloud by a line cloud by random generation of Collar Line Segments (CLS). While purely using position information might fall victim to correspondence ambiguity, additional information extracted from the point cloud like local plane structure [85], curvature & intensity [95], color information [86] are incorporated as multi-channel sources.

Once the data association established, tuning the outlier filter is a critical task for the success of the alignment. To filter out the outliers, each match is weighted by the inverse of the distance between the points involved in the matching process in [87]. In [88], outlier filtering is automatically tuned with respect to the overlap variations. Based on the true correspondence, many ICP variations use a closed form solution to iteratively compute the alignment based on cross-correlation of point-cloud [89]. To allow more generic minimization function other than simply Euclidean distance, nonlinear optimization framework is introduced to add additional constraints and dimensions based on $L2$ norm distance [90], such as more penalization along the planar patch direction [91] or edge direction [92]

2.2.2 Homogeneous Map Matching

Considering the efficiency and accuracy, matching the occupancy map is more suitable than raw sensor data and abstract topology. Firstly, a compact map is generated by compressing raw sensor data, as shown in Fig. 2.1, the large amount of data has been compressed about ninety percent. Secondly, although some information will lost when map is generated, the state-of-art three dimensional mapping algorithm would preserve the most of the sensor data, such as octomap shown in Fig. 2.1. In other words, sharing the occupancy/landmark maps [96] is preferred.

For 2D grid maps, topology-based approaches like hough transform [71] and Voronoi diagram [72] are applied to represent occupancy map on the abstract level. To calculate the transformation, a variety of standard topology map attributes are evaluated to compute the subgraph-isomorphisms in [75]. In [74], the author presents a hierarchical SLAM approach which uses the spectral information to register the local submaps. Besides, the probabilistic Voronoi matching is introduced to incorporate the motion uncertainty in [65]. For feature-based maps, Delaunay triangu-

(a) Raw sensor data with 6000 points (b) Octomap with 400 occupied voxels

Fig. 2.1 Raw sensor data of a plane and Octomap generated from the data lation based topology network is introduced to match the features in [97]. In [67], structural descriptors are extracted to match LIDAR point-cloud and sparse vision keypoint maps. Similarly, a histogram cross-correlation technique is proposed to perform map matching at both local and global levels in [76]. Aragues et al. [77] feature-based map merging with dynamic consensus on information increments.

With the increasing availability of high-quality 3D sensors, the dense registration algorithms have been employed to match maps composed of raw sensor data. In [64], probabilistic hypothesis density(PHD) SLAM is employed to match sparse features in simulation. CSHOT [98] descriptor and ICP [80] are integrated for submap-based map matching in [70]. For back-end optimization, [66] and [79] integrate raw sensor data to construct submaps and optimize constrained graph structures. However, aforementioned 3D map matching approaches are based on raw sensor data instead of probabilistic maps.

Since 3D occupancy grid map is generated by compressing the raw sensor data with a sensor measurement model [31], observation noise and motion uncertainty have been embedded into the map. The raw sensor data generated from the simulated environment are highly compressed to occupied voxels. Hence, the available characteristics of the occupancy map is modeled and combined into the registration process. The fusion of 3D volumetric maps generated using 3D probabilistic map is first proposed in [78], which utilizes ICP [80] to calculate the relative transformation by minimizing point-wise Euclidean distance. In [99], it is proven that registration algorithm combining invariant features will converge if the weight parameter is non-increasing. A recent paper [100] reviews the 3D scan matching methods up to date, concluding that Normal Distribution Transform (NDT) [93] provides most accurate results compared to others. NDT models the point cloud into a collection of Gaussian distributions with covariance and mean, and minimizes the sum of $L2$ norm. Since in our case the points are evenly distributed with a distance of the size of a voxel, the Gaussian distribution may not represent the environment explicitly compared to rich sensor data, so the error metric loses its original superiority. Based on this observation, probabilistic map information is incorporated to improve the accuracy of the estimated transformation in [69].

2.2.3 Heterogeneous Map Matching

For a group of robots in arbitrary (structured/semi-structured/unstructured) environments, single sensor is far from enough in mapping accuracy and robustness [101, 102]. Maps generated with heterogeneous sensors need to be matched to combine information with a higher information content [84] (see Fig. 2.2). Inspired by the probabilistic characteristic of expectation maximization(EM), several works consider more generalized scenario of correspondence between point sets with varying density. In [87], the author proposed an EM based ICP to improve the registration precision by considering the gaussian noise between the point sets. In [103], the author proposed a learning anisotropic ICP, which doesn't require manual tuning and update the parameters continually. In [84], the author further exploits the data association relationship and proposes an algorithm to align a sparse point cloud with a dense one, where each point is associated with a set of points. Then, EM based data association is extended to include semantic information and features, as well as internal measurements in [104].

For calibration on sensor level, research has been done in the field of extrinsic calibration between a 3D lidar, visual camera and thermal camera in the past decades. Zhang and Pless [105] first proposes to use checkerboard to calibrate a 2D lidar and a camera. Traditionally, one checkerboard should be put at different positions and orientations to collect multiple images, which is laborious. To alleviate the problem, [106] proposes to put several checkerboards at various positions and orientations at a time, so that taking only one shot is enough. They also make the calibration toolbox publicly available [107]. Recently, [108] proposes a novel checkerboard and is particularly suitable for the sparse 3D lidar. Following [108, 109] makes a progress by improving the detection accuracy of the circles. However, [109] only applies to a stereo camera. Meanwhile, a successful work on calibration of 3d-thermal-depth can be found in [110, 111]. However, the calibration on sensor level requires real time transferring raw sensor data among the robots, which is very challenging for real multiple robots applications.

In [84], the author only hard codes the K nearest neighborhood and does not discuss how to automatically tune the multiple data association strategy to balance the accuracy and efficiency. In [67], the author proposed a structural descriptors for matching point-cloud maps and vision keypoint maps. However, the paper presents a dense point-cloud map by rotating the 2D laser (Hokuyo) in off-line, 3D laser (Velodyne) usually tends to produce sparse data in real time. In addition, the structural descriptors are usually sensitive to parameters and requires fine tuning. In [63], the registration method is based on the matching of corresponding planar segments that

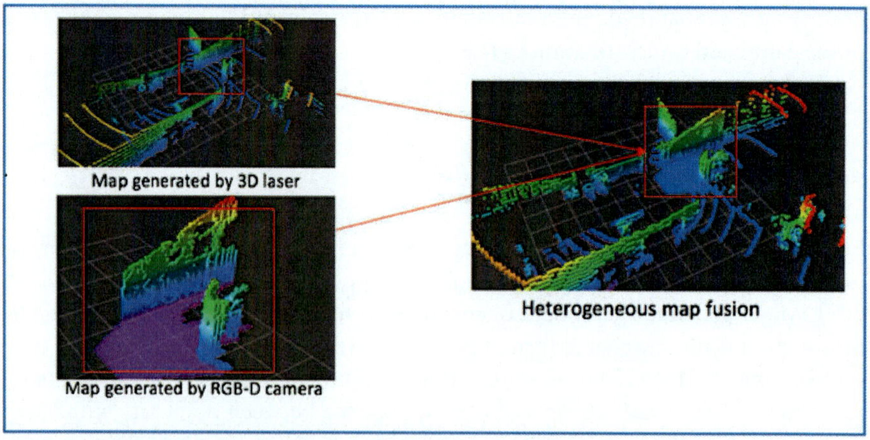

Fig. 2.2 The matching of maps generated by heterogeneous sensors

are extracted from the point clouds. An off-line approach for a globally optimization is also presented. In [112], the author compares the following descriptors a) local features: key-points or segments; b) descriptors: FPFH, SHOT, or ESF; and c) transformation estimations: RANSAC or FGR for registration of point clouds produced by Lidar and camera.

2.3 Collaborative Information Fusion

2.3.1 Map Inconsistency Detection

In registration algorithm, there is no guarantee that a global minimum will be found. A wrong transformation would lead to an inaccurate global map, which in turn is detrimental for robot exploration. Over the last decades, various SLAM approaches have been proposed to address the consistency problem. Especially for the Graph SLAM, which optimize the overall trajectories and sensor data to achieve a consistent global map. To increase the convergence of mapping, graph SLAM back-end has been proposed to minimize the accumulated error [113]. Based on that work, variants using dynamic covariance scaling [114] and switchable constraints [15] have been addressed.

However, it is still not yet possible to measure at what level the consistency of the map generated. In [115], an environment measurement model is proposed to verify the accuracy of transformation by calculating the percentage of inliers. In [83], the author analyses the sources of error for ICP and proposes a closed-form algorithm to estimate the covariance of ICP. In statistics, the paper models the histogram of the misalignment error metric into a truncated Gaussian distribution to evaluate the consistency in [116]. The research of identifying whether a map is consistent has not been extensively studied except the works proposed in [117–119]. In [117], the author applies a Bayesian representation of map posterior to revise the past wrong data association. Then it is extended to a multi-scan scenario with a cascaded map inconsistency statistical test in [118]. Based on that work, [119] developed a robot homing strategy whenever the map is detected to be inconsistent. These works pioneer the map inconsistency detection problem, and the use of this knowledge to improve autonomous mapping missions. However, these methods focus on sensor data level in a 2D environment with a single robot, which is infeasible to be extended to a multi-robot scenario in full 3D mapping missions.

Since it's reasonable to assume the submap is locally consistent [120], submap-based mapping strategy shows its superiority. In large environment, submap-based approach provides incremental [121] and efficient [122] approach for robot mapping. For multiple robots coordination [79], submap is even more suitable under the constraints of limited communication bandwidth and coverage. As submaps are generated over a short motion window of the robot, it is reasonable to assume that they are locally consistent [120]. To the extent of our knowledge, this is the first application of robust submap-based inconsistency detection for multi-robot mapping.

2.3.2 Probabilistic Information Integration

To distribute data among robots, a serial sequential strategy to communicate with robots is proposed in [123]. In DDF-SAM [52], a robust distributed communication framework is proposed, however solely provide simulations results under simplifying assumptions. In [124], the author proposed a distributed scheme for multiple data-gathering robots under limited communication and buffer size constraints. In [61], proposed an efficient and robust algorithm, an offboard ground station takes sensor data from all MAVs and fuses them into a detailed map. In [124], the author proposed a distributed scheme for multiple data-gathering robots under limited communication and buffer size constraints. In [125, 126], the author presents a calibration method based on raw sensor data.

When a robust and accurate transformation is calculated when the map is received, the source map is transformed to the coordination frame of the target map by a transformation function. Then, the key issue is to integrate the probabilistic information in partial maps into a global consistent map. In many of the previous methods, once the transformations between map coordinate frames are estimated, the final maps are generated by simply stitching the overlapped data [70] or averaging the occupancy probability value of corresponding pairs on voxel-wise level [78], where uncertainty propagation and merging is ignored. To measure the uncertainty of probabilistic map, Shannon entropy is used to measure the uncertainty of map in robot exploration in [28]. In [65], the paper introduces an entropy filter to reject fusion leading to a higher entropy, but is overconfident on the single map without taking both maps into consideration. Since the merging process should not introduce inconsistency for the global map, the Kullback-Leibler (KL) divergence based relative entropy filter is adopted to fuse occupancy probability consistently in [69].

In general, there are two types of uncertainty need to be integrated when performing map merging, which are uncertainty embedded in the map and uncertainty caused by the relative transformation. To propagate the transformation uncertainty, the covariance of transformation function is derived in [70]. In [83], the author proposed an closed-form estimate of ICP covariance. Then, the author calculates the covariance of transformed point by deriving the Jacobian of the transformation function in [65]. As for the map uncertainty, it can be represented by the occupancy probability in [31]. Given the uncertainty generated, the probability distribution should be merged to update the covariance. In [127], two 2D Gaussian distributions are multiplied and results in a compounded Gaussian distribution. Alternatively, the maps are fused on the voxel-wise level by directly taking the average of the probabilities in [78]. In [69], the relative transformation is evaluated based on a probabilistic measurement model and map dissimilarities are combined with relative entropy filter. Aforementioned approaches either assumes accurate transformation and only considers merging map uncertainty [69, 78, 127], or ignores the map uncertainty and focus on transformation uncertainty [65, 70]. In addition, none of those approaches tightly couple the map matching output with map merging, where two sub-problems are solved separately.

This part presents the relevant background of collaborative perception, localization and mapping and gives an extensive survey. The approaches of dense registration, map matching (homogeneous/heterogeneous), map sharing, inconsistency detection and information integration are carefully examined. Although, tremendous success have been made, a systematic collaborative perception, localization and mapping framework is still not studied in depth, especially in unstructured environments with heterogeneous sensors. Overall, most of the approaches just address the collaborative system from a specific viewpoint. However, they have not provided a comprehensive analysis of the collaborative perception, localization and mapping framework on scene understanding, feature extraction, map uncertainty modeling, probabilistic map matching, transformation evaluation and accurate map merging. By a close inspection it can be seen that the map matching approaches either falls to the scope of raw sensor data registration, or fails to model the nature of the 3D occupancy map. In addition, a full map merging process should consist of transformation evaluation, uncertainty derivation and information integration. Few of previous approaches address this issue in an adequate manner and most of the focus is given to integration of the information. The common thread of the following chapters is to develop a systematic collaborative perception, localization and mapping framework for autonomous robots that can address these challenges.

References

1. Smith RC, Cheeseman P (1986) On the representation and estimation of spatial uncertainty. Int J Robot Res 5(4):56–68
2. Julier SJ, Uhlmann JK (1997) New extension of the kalman filter to nonlinear systems. In: AeroSense'97, international society for optics and photonics, pp 182–193
3. Dissanayake MWMG, Newman P, Clark S, Durrant-Whyte HF, Csorba M (2001) A solution to the simultaneous localization and map building (slam) problem. IEEE Trans Robot Autom 17(3):229–241
4. Huang S, Dissanayake G (2007) Convergence and consistency analysis for Kalman filter based SLAM. IEEE Trans Robot 23(5):1036–1049
5. Montemerlo M, Thrun S, Koller D, Wegbreit B et al (2002) Fastslam: a factored solution to the simultaneous localization and mapping problem. In: AAAI/IAAI, pp 593–598
6. Montemerlo M, Thrun S (2007) Fastslam 2.0. FastSLAM: a scalable method for the simultaneous localization and mapping problem in robotics, pp 63–90
7. Thrun S, Liu Y, Koller D, Ng AY, Ghahramani Z, Durrant-Whyte H (2004) Simultaneous localization and mapping with sparse extended information filters. Int J Robot Res 23(7–8):693–716
8. Thrun S, Burgard W, Fox D (2005) Probabilistic robotics. MIT Press
9. Nüchter A, Lingemann K, Hertzberg J, Surmann H (2005) 6d slam with approximate data association. In: 12th international conference on advanced robotics, 2005. ICAR'05. Proceedings. IEEE, pp 242–249
10. Ng PC, Henikoff S (2003) Sift: predicting amino acid changes that affect protein function. Nucleic Acids Res 31(13):3812–3814
11. Mullane J, Vo BN, Adams MD, Vo BT (2011) A random-finite-set approach to Bayesian SLAM. IEEE Trans Robot 27(2):268–282
12. Thrun S, Montemerlo M (2006) The graph slam algorithm with applications to large-scale mapping of urban structures. Int J Robot Res 25(5–6):403–429

13. Kaess M, Ranganathan A, Dellaert F (2008) iSAM: incremental smoothing and mapping. IEEE Trans Robot 24(6):1365–1378
14. Kümmerle R, Steder B, Dornhege C, Kleiner A, Grisetti G, Burgard W (2011) Large scale graph-based slam using aerial images as prior information. Auton Robots 30(1):25–39
15. Sünderhauf N, Protzel P (2012) Switchable constraints for robust pose graph slam. In: 2012 IEEE/RSJ international conference on intelligent robots and systems, pp 1879–1884
16. Latif Y, Cadena C, Neira J (2013) Robust loop closing over time for pose graph slam. The Int J Robot Res 0278364913498910
17. Kümmerle R, Grisetti G, Strasdat H, Konolige K, Burgard W (2011) g 2 o: a general framework for graph optimization. In: 2011 IEEE international conference on robotics and automation (ICRA), IEEE, pp 3607–3613
18. Thrun S, Burgard W, Fox D (2005) Probabilistic robotics (intelligent robotics and autonomous agents). The MIT Press
19. Strasdat H, Davison AJ, Montiel JMM, Konolige K (2011) Double window optimisation for constant time visual slam. In: 2011 international conference on computer vision, pp 2352–2359. https://doi.org/10.1109/ICCV.2011.6126517
20. Fanfani M, Bellavia F, Colombo C (2016) Accurate keyframe selection and keypoint tracking for robust visual odometry. Mach Vis Appl 27(6):833–844. https://doi.org/10.1007/s00138-016-0793-3
21. Lim H, Lim J, Kim HJ (2014) Real-time 6-dof monocular visual slam in a large-scale environment. In: 2014 IEEE international conference on robotics and automation (ICRA), pp 1532–1539. https://doi.org/10.1109/ICRA.2014.6907055
22. Strasdat H, Montiel J, Davison AJ (2012) Visual slam: why filter? Image Vis Comput 30(2):65–77. https://doi.org/10.1016/j.imavis.2012.02.009, http://www.sciencedirect.com/science/article/pii/S0262885612000248
23. Lecun Y, Bottou L, Bengio Y, Haffner P (1998) Gradient-based learning applied to document recognition. Proc IEEE 86(11):2278–2324
24. Badrinarayanan V, Kendall A, Cipolla R (2017) Segnet: a deep convolutional encoder-decoder architecture for image segmentation. IEEE Trans Pattern Anal Mach Intell 39(12):2481–2495
25. Lin G, Milan A, Shen C, Reid I (2017) Refinenet: multi-path refinement networks for high-resolution semantic segmentation. In: Proceedings of the IEEE conference on computer vision and pattern recognition, pp 1925–1934
26. Chen LC, Papandreou G, Schroff F, Adam H (2017) Rethinking atrous convolution for semantic image segmentation. arXiv preprint arXiv:170605587
27. Wang K, Liu YH, Li L (2014) A simple and parallel algorithm for real-time robot localization by fusing monocular vision and odometry/ahrs sensors. IEEE/ASME Trans Mechatron 19(4):1447–1457
28. Carrillo H, Dames P, Kumar V, Castellanos JA (2015) Autonomous robotic exploration using occupancy grid maps and graph slam based on shannon and renyi entropy. In: 2015 IEEE international conference on robotics and automation (ICRA), pp 487–494
29. Kostavelis I, Gasteratos A (2015) Semantic mapping for mobile robotics tasks: a survey. Robot Auton Syst 66:86–103
30. Galindo C, Saffiotti A, Coradeschi S, Buschka P, Fernandez-Madrigal JA, González J (2005) Multi-hierarchical semantic maps for mobile robotics. In: 2005 IEEE/RSJ international conference on intelligent robots and systems, IEEE, pp 2278–2283
31. Hornung A, Wurm KM, Bennewitz M, Stachniss C, Burgard W (2013) OctoMap: an efficient probabilistic 3d mapping framework based on octrees. Auton Robots
32. Yang S, Huang Y, Scherer S (2017) Semantic 3d occupancy mapping through efficient high order crfs. In: 2017 IEEE/RSJ international conference on intelligent robots and systems (IROS), IEEE, pp 590–597
33. Berrio JS, Zhou W, Ward J, Worrall S, Nebot E (2018) Octree map based on sparse point cloud and heuristic probability distribution for labeled images. In: 2018 IEEE/RSJ international conference on intelligent robots and systems (IROS), IEEE, pp 3174–3181

34. Bayram H, Bozma HI (2010) Multi-robot navigation with limited communication-deterministic vs game-theoretic networks. In: 2010 IEEE/RSJ international conference on intelligent robots and systems, IEEE, pp 1825–1830
35. Kostavelis I, Charalampous K, Gasteratos A, Tsotsos JK (2016) Robot navigation via spatial and temporal coherent semantic maps. Eng Appl Artif Intell 48:173–187
36. Yu H, Moon J, Lee B (2019) A variational observation model of 3d object for probabilistic semantic slam. In: 2019 international conference on robotics and automation (ICRA), IEEE, pp 5866–5872
37. Bârsan IA, Liu P, Pollefeys M, Geiger A (2018) Robust dense mapping for large-scale dynamic environments. In: 2018 IEEE international conference on robotics and automation (ICRA), IEEE, pp 7510–7517
38. Kochanov D, Ošep A, Stückler J, Leibe B (2016) Scene flow propagation for semantic mapping and object discovery in dynamic street scenes. In: 2016 IEEE/RSJ international conference on intelligent robots and systems (IROS), IEEE, pp 1785–1792
39. Michael N, Shen S, Mohta K, Mulgaonkar Y et al (2012) Collaborative mapping of an earthquake-damaged building via ground and aerial robots. J Field Robot 29(5):832–841
40. Hussein M, Renner M, Watanabe M, Iagnemma K (2013) Matching of ground-based lidar and aerial image data for mobile robot localization in densely forested environments. In: 2013 IEEE/RSJ international conference on intelligent robots and systems (IROS), IEEE, pp 1432–1437
41. Trautman P, Ma J, Murray RM, Krause A (2015) Robot navigation in dense human crowds: statistical models and experimental studies of human–robot cooperation. Int J Robot Res 34(3):335–356
42. Fox D, Ko J, Konolige K, Limketkai B, Schulz D, Stewart B (2006) Distributed multirobot exploration and mapping. Proc IEEE 94(7):1325–1339
43. Elibol A, Kim J, Gracias N, Garcia R (2014) Efficient image mosaicing for multi-robot visual underwater mapping. Pattern Recognit Lett 46:20–26
44. Fenwick JW, Newman PM, Leonard JJ (2002) Cooperative concurrent mapping and localization. In: IEEE international conference on robotics and automation, 2002. Proceedings. ICRA'02. IEEE, vol 2, pp 1810–1817
45. Mourikis AI, Roumeliotis SI (2006) Predicting the performance of cooperative simultaneous localization and mapping (c-slam). Int J Robot Res 25(12):1273–1286. https://doi.org/10.1177/0278364906072515
46. Howard A (2006) Multi-robot simultaneous localization and mapping using particle filters. Int J Robot Res 25(12):1243–1256
47. Carlone L, Ng MK, Du J, Bona B, Indri M (2010) Rao-blackwellized particle filters multi robot slam with unknown initial correspondences and limited communication. In: 2010 IEEE international conference on robotics and automation, pp 243–249. https://doi.org/10.1109/ROBOT.2010.5509307
48. Andersson LA, Nygards J (2008) C-sam: multi-robot slam using square root information smoothing. In: IEEE international conference on robotics and automation, 2008. ICRA 2008. IEEE, pp 2798–2805
49. Burgard W, Moors M, Stachniss C, Schneider FE (2005) Coordinated multi-robot exploration. IEEE Trans Robot 21(3):376–386
50. Saeedi S, Paull L, Trentini M, Li H (2011) Neural network-based multiple robot simultaneous localization and mapping. IEEE Trans Neural Netw 22(12):2376–2387
51. Moratuwage D, Wang D, Rao A, Senarathne N, Wang H (2014) Rfs collaborative multivehicle slam: slam in dynamic high-clutter environments. IEEE Robot Autom Mag 21(2):53–59
52. Cunningham A, Indelman V, Dellaert F (2013) Ddf-sam 2.0: consistent distributed smoothing and mapping. In: 2013 IEEE international conference on robotics and automation, pp 5220–5227. https://doi.org/10.1109/ICRA.2013.6631323
53. Dong J, Nelson E, Indelman V, Michael N, Dellaert F (2015) Distributed real-time cooperative localization and mapping using an uncertainty-aware expectation maximization approach. In: 2015 IEEE international conference on robotics and automation (ICRA), pp 5807–5814

54. Zhou XS, Roumeliotis SI (2008) Robot-to-robot relative pose estimation from range measurements. IEEE Trans Robot 24(6):1379–1393
55. Kim B, Kaess M, Fletcher L, Leonard J, Bachrach A, Roy N, Teller S (2010) Multiple relative pose graphs for robust cooperative mapping. In: 2010 IEEE international conference on robotics and automation (ICRA), IEEE, pp 3185–3192
56. Dinnissen P, Givigi Jr SN, Schwartz HM (2012) Map merging of multi-robot slam using reinforcement learning. In: 2012 IEEE international conference on systems, man, and cybernetics (SMC), IEEE, pp 53–60
57. Konolige K, Marder-Eppstein E, Marthi B (2011) Navigation in hybrid metric-topological maps. In: 2011 IEEE international conference on robotics and automation (ICRA), pp 3041–3047. https://doi.org/10.1109/ICRA.2011.5980074
58. Moreno FA, Blanco JL, Gonzalez-Jimenez J (2015) A constant-time SLAM back-end in the continuum between global mapping and submapping: application to visual stereo SLAM. Int J Robot Res 0278364915619238
59. Angeli A, Filliat D, Doncieux S, Meyer JA (2008) Fast and incremental method for loop-closure detection using bags of visual words. IEEE Trans Robot 24(5):1027–1037
60. Saeedi S, Trentini M, Seto M, Li H (2016) Multiple-robot simultaneous localization and mapping: a review. J Field Robot 33(1):3–46
61. Scaramuzza D, Achtelik MC, Doitsidis L, Friedrich F, Kosmatopoulos E (2014) Vision-controlled micro flying robots: from system design to autonomous navigation and mapping in GPS-denied environments. IEEE Robot Autom Mag 21(3):26–40
62. Rashid AT, Frasca M, Ali AA (2015) Multi-robot localization and orientation estimation using robotic cluster matching algorithm. Robot Auton Syst 63:108–121
63. Surmann H, Berninger N, Worst R (2017) 3d mapping for multi hybrid robot cooperation. In: 2017 IEEE/RSJ international conference on intelligent robots and systems (IROS), pp 626–633. https://doi.org/10.1109/IROS.2017.8202217
64. Niedfeldt PC, Speranzon A, Surana A (2015) Distributed map fusion with sporadic updates for large domains. In: 2015 IEEE international conference on robotics and automation, pp 2806–2813
65. Saeedi S, Paull L, Trentini M, Seto M (2014) Group mapping: a topological approach to map merging for multiple robots. Robot Autom Mag IEEE 21(2):60–72
66. Bonanni TM, Corte BD, Grisetti G (2017) 3-D map merging on pose graphs. IEEE Robot Autom Lett 2(2):1031–1038. https://doi.org/10.1109/LRA.2017.2655139
67. Gawel A, Cieslewski T, Dub R, Bosse M, Siegwart R (2016) Structure-based vision-laser matching. In: 2016 IEEE/RSJ international conference on intelligent robots and systems (IROS), pp 182–188. https://doi.org/10.1109/IROS.2016.7759053
68. Jessup J, Givigi SN, Beaulieu A (2017) Robust and efficient multirobot 3-d mapping merging with octree-based occupancy grids. IEEE Syst J 11(3):1723–1732
69. Yue Y, Wang D, Senarathne P, Moratuwage D (2016) A hybrid probabilistic and point set registration approach for fusion of 3D occupancy grid maps. In: 2016 IEEE international conference on systems, man, and cybernetics, pp 1975–1980
70. Brand C, Schu MJ, Hirsch H (2015) Submap matching for stereo-vision based indoor/outdoor SLAM. In: IEEE/RSJ international conference on intelligent robots and systems, pp 5670–5677
71. Carpin S (2008) Fast and accurate map merging for multi-robot systems. Auton Robots 25(3):305–316
72. Saeedi S, Paull L, Trentini M (2014) Map merging for multiple robots using Hough peak matching. Robot Auton Syst 62(10):1408–1424
73. Saarinen J, Andreasson H, Stoyanov T, Lilienthal AJ (2013) Normal distributions transform Monte-Carlo localization (NDT-MCL). In: 2013 IEEE/RSJ international conference on intelligent robots and systems, pp 382–389. https://doi.org/10.1109/IROS.2013.6696380
74. Oberlünd'er J, Roennau A, Dillmann R (2013) Hierarchical SLAM using spectral submap matching with opportunities for long-term operation. In: 2013 16th international conference on advanced robotics (ICAR), pp 1–7

75. Schwertfeger S, Birk A (2016) Map evaluation using matched topology graphs. Auton Robots 40(5):761–787
76. Bosse M, Zlot R (2008) Map matching and data association for large-scale two-dimensional laser scan-based slam. Int J Robot Res 27(6):667–691. https://doi.org/10.1177/0278364908091366
77. Aragues R, Sagues C, Mezouar Y (2015) Feature-based map merging with dynamic consensus on information increments. Auton Robots 38(3):243–259
78. Jessup J, Givigi SN, Beaulieu A (2015) Robust and efficient multirobot 3-d mapping merging with octree-based occupancy grids. IEEE Syst J PP(99):1–10
79. Schuster MJ, Brand C, Hirschmüller H, Suppa M, Beetz M (2015) Multi-robot 6D graph SLAM connecting decoupled local reference filters. In: 2015 IEEE/RSJ international conference on intelligent robots and systems (IROS), IEEE, pp 5093–5100
80. Besl PJ, McKay ND (1992) A method for registration of 3-d shapes. IEEE Trans Pattern Anal Mach Intell 14(2):239–256
81. Taguchi Y, Jian YD, Ramalingam S, Feng C (2013) Point-plane slam for hand-held 3d sensors. In: 2013 IEEE international conference on robotics and automation, pp 5182–5189. https://doi.org/10.1109/ICRA.2013.6631318
82. Das A, Servos J, Waslander SL (2013) 3d scan registration using the normal distributions transform with ground segmentation and point cloud clustering. In: 2013 IEEE international conference on robotics and automation, pp 2207–2212. https://doi.org/10.1109/ICRA.2013.6630874
83. Censi A (2007) An accurate closed-form estimate of ICP's covariance. In: Proceedings 2007 IEEE international conference on robotics and automation, pp 3167–3172. https://doi.org/10.1109/ROBOT.2007.363961
84. Agamennoni G, Fontana S, Siegwart RY, Sorrenti DG (2016) Point clouds registration with probabilistic data association. In: 2016 IEEE/RSJ international conference on intelligent robots and systems (IROS), pp 4092–4098. https://doi.org/10.1109/IROS.2016.7759602
85. Segal A, Haehnel D, Thrun S (2009) Generalized-ICP. In: Robotics: science and systems, vol 2, p 435
86. Servos J, Waslander SL (2017) Multi-channel generalized-ICP: a robust framework for multi-channel scan registration. Robot Auton Syst 87(Supplement C):247–257
87. Liu J, Du S, Zhang C, Zhu J (2014) Probability iterative closest point algorithm for position estimation. In: International IEEE conference on intelligent transportation systems, pp 458–463. https://doi.org/10.1109/ITSC.2014.6957732
88. Nobili S, Scona R, Caravagna M (2017) Overlap-based ICP tuning for robust localization of a humanoid robot. In: IEEE international conference on robotics and automation, pp 4721–4728. https://doi.org/10.1109/ICRA.2017.7989547
89. Ying S, Peng J, Du S, Qiao H (2009) A scale stretch method based on ICP for 3d data registration. IEEE Trans Autom Sci Eng 6(3):559–565. https://doi.org/10.1109/TASE.2009.2021337
90. Li Q, Xiong R, Vidal-Calleja T (2017) A GMM based uncertainty model for point clouds registration. Robot Auton Syst 91(Supplement C):349–362
91. Viejo D, Cazorla M (2014) A robust and fast method for 6DoF motion estimation from generalized 3D data. Auton Robots 36(4):295–308
92. Serafin J, Grisetti G (2017) Using extended measurements and scene merging for efficient and robust point cloud registration. Robot Auton Syst 92(Supplement C):91–106
93. Stoyanov TD, Magnusson M, Andreasson H, Lilienthal A (2012) Fast and accurate scan registration through minimization of the distance between compact 3D NDT representations. Int J Robot Res 31(12):1377–1393
94. Velas M, Spanel M, Herout A (2016) Collar line segments for fast odometry estimation from velodyne point clouds. In: IEEE international conference on robotics and automation, pp 4486–4495. https://doi.org/10.1109/ICRA.2016.7487648
95. He Y, Liang B, Yang J, Li S, He J (2017) An iterative closest points algorithm for registration of 3d laser scanner point clouds with geometric features. Sensors 17(8)

96. Yue Y, Senarathne PN, Yang C, Zhang J, Wen M, Wang D (2018) Hierarchical probabilistic fusion framework for matching and merging of 3-d occupancy maps. IEEE Sens J 18(21):8933–8949
97. Cunningham A, Wurm KM, Burgard W, Dellaert F (2012) Fully distributed scalable smoothing and mapping with robust multi-robot data association. In: 2012 IEEE international conference on robotics and automation (ICRA), pp 1093–1100
98. Tombari F, Salti S, Stefano LD (2011) A combined texture-shape descriptor for enhanced 3d feature matching. In: 2011 18th IEEE international conference on image processing, pp 809–812. https://doi.org/10.1109/ICIP.2011.6116679
99. Sharp GC, Lee SW, Wehe DK (2002) ICP registration using invariant features. IEEE Trans Pattern Anal Mach Intell 24(1):90–102
100. Magnusson M, Vaskevicius N, Stoyanov T, Pathak K, Birk A (2015) Beyond points: evaluating recent 3d scan-matching algorithms. In: 2015 IEEE international conference on robotics and automation (ICRA), pp 3631–3637
101. Yue Y, Yang C, Zhang J, Wen M, Wu Z, Zhang H, Wang D (2020) Day and night collaborative dynamic mapping in unstructured environment based on multimodal sensors. In: 2020 IEEE international conference on robotics and automation (ICRA)
102. Yue Y, Yang C, Wang Y, Senarathne PCN, Zhang J, Wen M, Wang D (2019) A multilevel fusion system for multirobot 3-d mapping using heterogeneous sensors. IEEE Syst J 14(1):1341–1352
103. Lee B, Lee DD (2016) Learning anisotropic ICP (LA-ICP) for robust and efficient 3D registration. In: 2016 IEEE international conference on robotics and automation (ICRA), pp 5040–5045. https://doi.org/10.1109/ICRA.2016.7487709
104. Bowman SL, Atanasov N, Daniilidis K, Pappas GJ (2017) Probabilistic data association for semantic slam. In: 2017 IEEE international conference on robotics and automation (ICRA), vol 2, pp 1092–1098
105. Zhang Q, Pless R (2004) Extrinsic calibration of a camera and laser range finder (improves camera calibration). In: 2004 IEEE/RSJ international conference on intelligent robots and systems (IROS) (IEEE Cat. No.04CH37566), vol 3, pp 2301–2306
106. Geiger A, Moosmann F, Car Ö, Schuster B (2012) Automatic camera and range sensor calibration using a single shot. In: 2012 IEEE international conference on robotics and automation, pp 3936–3943
107. Geiger A, Moosmann F, Car Ö, Schuster B (2012) Camera and range sensor calibration toolbox. http://www.cvlibs.net/software/calibration/index.php
108. Velas M, Spanel M, Materna Z, Herout A (2014) Calibration of RGB camera with velodyne lidar. In: WSCG 2014
109. Guindel C, Beltrán J, Martín D, García F (2017) Automatic extrinsic calibration for lidar-stereo vehicle sensor setups. In: 2017 IEEE 20th international conference on intelligent transportation systems (ITSC), pp 1–6
110. Vidas S, Moghadam P, Bosse M (2013) 3d thermal mapping of building interiors using an RGB-D and thermal camera. In: 2013 IEEE international conference on robotics and automation, pp 2311–2318
111. Vidas S, Lakemond R, Denman S, Fookes C, Sridharan S, Wark T (2012) A mask-based approach for the geometric calibration of thermal-infrared cameras. IEEE Trans Instrum Meas 61(6):1625–1635
112. Gawel A, Dubé R, Surmann H, Nieto J, Siegwart R, Cadena C (2017) 3d registration of aerial and ground robots for disaster response: an evaluation of features, descriptors, and transformation estimation. In: 2017 IEEE international symposium on safety, security and rescue robotics (SSRR), pp 27–34. https://doi.org/10.1109/SSRR.2017.8088136
113. Grisetti G, Kummerle R, Stachniss C, Burgard W (2010) A tutorial on graph-based SLAM. IEEE Intell Transp Syst Mag 2(4):31–43
114. Agarwal P, Tipaldi GD, Spinello L (2013) Robust map optimization using dynamic covariance scaling. In: 2013 IEEE international conference on robotics and automation, pp 62–69. https://doi.org/10.1109/ICRA.2013.6630557

115. Endres F, Hess J, Sturm J, Cremers D, Burgard W (2014) 3-D mapping with an RGB-D camera. IEEE Trans Robot 30(1):177–187
116. Yue Y, Wang D, Sena PGCN, Yang C (2017) Robust submap-based probabilistic inconsistency detection for multi-robot mapping. In: 2017 European conference on mobile robots (ECMR), pp 1–6
117. Hähnel D, Thrun S, Wegbreit B, Burgard W (2005) Towards lazy data association in SLAM. In: Robotics Research. The Eleventh International Symposium, pp 421–431. Springer
118. Mazuran M, Diego Tipaldi G, Spinello L, Burgard W, Stachniss C (2014) A statistical measure for map consistency in slam. In: 2014 IEEE international conference on robotics and automation (ICRA), pp 3650–3655
119. Bogoslavskyi I, Mazuran M, Stachniss C (2016) Robust homing for autonomous robots. In: 2016 IEEE international conference on robotics and automation (ICRA), pp 2550–2556. https://doi.org/10.1109/ICRA.2016.7487410
120. Rehder E, Albrecht A (2015) Submap-based SLAM for road markings. In: 2015 IEEE intelligent vehicles symposium (IV), pp 1393–1398
121. Blanco JL, Fernández-Madrigal JA, Gonzez J (2008) Toward a unified Bayesian approach to hybrid metric-topological SLAM. IEEE Trans Robot 24(2):259–270. https://doi.org/10.1109/TRO.2008.918049
122. Ni K, Dellaert F (2010) Multi-level submap based slam using nested dissection. In: 2010 IEEE/RSJ international conference on intelligent robots and systems (IROS), IEEE, pp 2558–2565
123. Yue Y, Zhao C, Li R, Yang C, Zhang J, Wen M, Wang Y, Wang D (2020) A hierarchical framework for collaborative probabilistic semantic mapping. In: 2020 IEEE international conference on robotics and automation (ICRA),
124. Guo M, Zavlanos MM (2018) Multirobot data gathering under buffer constraints and intermittent communication. IEEE Trans Robot 34(4):1082–1097. https://doi.org/10.1109/TRO.2018.2830370
125. Levinson J, Thrun S (2013) Automatic online calibration of cameras and lasers. In: Robotics: science and systems, pp 24–28
126. Maddern W, Newman P (2016) Real-time probabilistic fusion of sparse 3d lidar and dense stereo. In: 2016 IEEE/RSJ international conference on intelligent robots and systems (IROS), pp 2181–2188
127. Stroupe AW, Martin MC, Balch T (2001) Distributed sensor fusion for object position estimation by multi-robot systems. In: Proceedings 2001 ICRA. IEEE international conference on robotics and automation (Cat. No.01CH37164), vol 2, pp 1092–1098

Chapter 3
Point Registration Approach for Map Fusion

One of the major challenges in multi-robot collaboration is to fuse the partial maps generated by individual robots into a consistent global map. We address 3D volumetric map fusion by extending the well known iterative closest point (ICP) algorithm to include probabilistic distance and surface information. In addition, the relative transformation is evaluated based on Mahalanobis distance and map dissimilarities are integrated using relative entropy filter. The efficiency of the proposed algorithm is evaluated using maps generated from both simulated and real environments and is shown to generate more consistent global maps.

3.1 Introduction

With the continuing development in the field of mobile robotics and SLAM, in particular, the use of mobile robots to autonomously explore and map an environment using multiple robots has caught the attention of many robotics researchers and practitioners. Coordination among multiple robots enable them to perform difficult tasks in challenging environments more efficiently and reliably, such as search and rescue missions.

One of the major challenges in such missions is to fuse the data generated by individual robots, which can be performed at different levels. Under the condition of limited communication and memory, a compact map generated by compressing raw sensor data is preferred. Hence, it's vital to fuse the partial maps generated by individual robots into a more consistent global map, which is further compounded with 3D volumetric maps where the map fusion algorithms have to process large amounts of 3D map data.

Y. Yue and D. Wang, *Collaborative Perception, Localization and Mapping for Autonomous Systems*, Springer Tracts in Autonomous Systems 2,
https://doi.org/10.1007/978-981-15-8860-0_3

The research in map fusion has not been studied extensively, especially for 3D volumetric maps. Majority of the existing approaches focus on feature matching and topology matching for 2D maps. As the availability of high-quality 3D sensors increase, Octomap based on Octree data structure has been widely used to model the environment with a full probabilistic 3D model. In order to get an accurate and systematic map fusion algorithm, two requirements are set.

Relative Transformation: Since the book is working on the compressed map level, all the characteristics of the map should be utilized to compute an accurate transformation. In addition, point-wise correspondence should be provided for probability fusion.

Probability Fusion: Many of the existing researches pay much attention to compute relative transformation, however place less emphasis on how to fuse the maps. Due to different view and SLAM error, individual maps will have varying probability even for the same object. To fuse the maps into a global one, dissimilarities between individual maps should be measured to fuse the maps and decrease the total uncertainty.

This chapter addresses the above requirements and presents a novel method with the following contributions:

Firstly, probabilistic and surface information are incorporated in the registration algorithm, which is called Occupancy Iterative Closet Point (OICP). OICP combines the strength of ICP with the associated probabilistic and surface information of 3D occupancy map, which tends to match voxels with similar occupancy probability.

Secondly, an environment measurement model (EMM) is applied to decide whether to accept or reject the computed relative transformation. Then a relative entropy filter is proposed to fuse the map by combing occupancy probability from both maps which also decreases the overall global map uncertainty.

The chapter is organized as follows: Sect. 3.2 introduces the Occupancy Iterative Closet Point algorithm. Section 3.3 details the evaluation of the computed relative transformation and the fusion of probabilities. Section 3.4 presents the outcome of the experiments conducted using simulated and real data. Section 3.5 concludes the chapter and discusses the future work.

3.2 OICP Algorithm

In this section, occupancy iterative closet point (OICP) is presented to incorporate occupancy probability and surface information of 3D occupancy map, which computes the relative transformation between partial maps under the condition of given consistent partial occupancy maps and initial position estimation.

Figure 3.1 illustrates an example of Octomap and the associated point cloud extracted from the map. Octomap is a three-dimensional voxel map based on octrees, providing a volumetric representation of space which is a compact, probabilistic, efficient model for a variety of robotic applications including flying robots and ground

(a) Octomap generated from a corridor (b) Point cloud of voxel centers

Fig. 3.1 An example of Octomap and its related point cloud

robots. The measurements are integrated using a sensor model to provide the probability of grid. Then a grid cell probability is updated using a Bayesian process.

Given an Octomap M_r, the point cloud m_r extracted from Octomap is defined as

$$m_r = (m_r^1, m_r^2, \cdots, m_r^i, \cdots)$$

where $(m_{rx}^i, m_{ry}^i, m_{rz}^i, L_{m_r^i}, d_{m_r})$ is a tuple that includes

$$m_r = \begin{cases} (m_{rx}^i, m_{ry}^i, m_{rz}^i) & \text{position of the extracted voxel center} \\ L_{m_r^i} & \text{the occupancy probability log-likelihood} \\ d_{m_r} & \text{The size of the voxel in Octree} \end{cases}$$

In the registration process, the source point cloud will be transformed by matrix T iteratively to minimize the error metric with target point cloud, while target point cloud remains fixed. Here, point cloud m_r is assigned as target point cloud and m_s as source point cloud. The transformation matrix T, consisting of rotation matrix R and translation matrix t. The transformation matrix T can be written as the form of affine matrices in Eq. (3.1).

$$T = \begin{bmatrix} R_{3\times3} & t_{3\times1} \\ 0 & I_{1\times1} \end{bmatrix} \tag{3.1}$$

The OICP can be divided into two stages. For the first stage, point-to-plane ICP is applied to make use of the smooth surface information of Octomap using Eq. (3.2), where η is the surface normal of target point cloud m_r.

$$f = \min_T \sum_{i,j} ||(Rm_s^j + t - m_r^i) \cdot \eta||^2 \tag{3.2}$$

In the second stage, voxel occupancy probability is incorporated in order to achieve accurate transformation with point-wise correspondence. The iterative registration using invariant features is proposed in [1], it is proven that a weighted linear combination of positional and feature distances will converge if the weight parameter is non-increasing. The work is proved in a common framework. Specifically, at each time step of map fusion, the occupancy is regarded as an invariant feature. Hence, the weighted linear combination is utilized to combine positional distance and feature distance into a single point-wise distance metric, where each point is represented by its three positional coordinates and the attached one dimensional occupancy probability coordinate. Points are matched in this 4 dimensional space.

The point-wise distance is described by the summation of Euclidean distance $d_e(m_r^i, m_s^j)$ and feature distance $d_o(p_{m_r^i}, p_{m_s^j})$. The weight parameter ω_o is defined to balance the contribution between positions and features. The general form for the distance metric is defined in Eq. (3.3).

$$d = d_e(m_r^i, m_s^j) + \omega_o d_o(p_{m_r^i}, p_{m_s^j}) \tag{3.3}$$

For Euclidean distance, the classical ICP is applied to calculate the relative transformation by minimizing point-wise Euclidean distance using Eq. (3.4). ICP iteratively revises the transformation needed to minimize the distance from the source point cloud m to the target point cloud n. ICP will be stop when a defined threshold is achieved, for example, the maximum number of iterations.

$$d_e = ||Rm_s^j + t - m_r^i||^2 \tag{3.4}$$

For feature distance, the occupancy probability is set as an invariant feature. The probability of voxel m_r^i given observation $z_{1:t}^i$ is updated using log-odds notation in Octomap. Although occupancy probability will recursively update when new sensor measurements are observed, it can be viewed as invariant features at each time step.

$$p(m_r^i|z_{1:t}^i) = \left(1 + \frac{1 - p(m_r^i|z_t^i)}{p(m_r^i|z_t^i)} \frac{1 - p(m_r^i|z_{1:t-1}^i)}{p(m_r^i|z_{1:t-1}^i)} \frac{p(m_r^i)}{1 - p(m_r^i)}\right)^{-1} \tag{3.5}$$

Using log-odds notation, Eq. (3.5) can be simplified to:

$$L(m_r^i|z_{1:t}^i) = L(m_r^i|z_{1:t-1}^i) + L(m_r^i|z_t{}^i) \tag{3.6}$$

Note that log-odds can be easily converted to probabilities using Eq. (3.7).

$$L(m_r^i|z_{1:t}^i) = \log \frac{p(m_r^i|z_{1:t}^i)}{1 - p(m_r^i|z_{1:t}{}^i)} \tag{3.7}$$

Then, the feature distance is defined as:

$$d_o = ||p_{m_r^i} - p_{m_s^j}||^2 \tag{3.8}$$

Combining Euclidean distance and feature distance, occupancy iterative closet point is formulated in Eq. (3.9). Here the Euclidean distance is defined as the distance between center points of the voxel, and feature distance is represented by the difference in associated occupancy probabilities.

$$f_{oicp} = \min_T \sum_{i,j} ||Rm_s^j + t - m_r^i||^2 + \omega_o||p_{m_s^j} - p_{m_r^i}||^2 \tag{3.9}$$

The OICP iteratively computes the transformation which minimizes the proposed error metric between corresponding points. And the trade off between positional and feature value is computed by weight parameter ω_o. The weight parameter is a scale factor to determine the contribution of positional and probability value. To model this weight, the uncertainty of positional error caused by mismatching and probability error caused by sensor noised are estimated. The OICP algorithm can be summarized as follows:

3.2.1 Uncertainty in Occupancy Probability

As Octomap is generated using noisy pose estimates provided by SLAM outputs and noisy sensor data, there is always an inherent level of uncertainty associated with them. When the uncertainty of the map is high, less confident is given on the map. On the contrary, when the uncertainty of the map is low, more weight is assigned to the map. Therefore, Shannon entropy is introduced to measure the uncertainty of the probabilistic voxel m_r^i using Eq. (3.10). $H(m_r^i)$ is defined to represent the entropy of probabilistic voxel m_r^i.

$$H(m_r^i) = -p_{m_r^i} \log_2 p_{m_r^i} \tag{3.10}$$

The occupancy probability of $p_{m_r^i}$ is within the range of [0, 1], where

$$p_{m_r^i} \begin{cases} \in [0, 0.5) & \text{the voxel is free} \\ = 0.5 & \text{unknown} \\ \in (0.5, 1] & \text{the voxel the occupied} \end{cases}$$

The higher the probability, the more certainty the voxel is occupied. The lower probability, the more certainty the voxel is free. Here, the situation of voxel is occupied, which means $p_{m_r^i} \in (0.5, 1]$. The value of $H(m_r^i)$ with the change of $p_{m_r^i} \in (0.5, 1]$ is shown in Fig. 3.2, which is a monotonic decreasing function. More specifically,

$$\begin{cases} H(m_r^i)_{min} = 0 & \text{when } p_{m_r^i} = 1 \\ H(m_r^i)_{max} = 0.5 & \text{when } p_{m_r^i} = 0.5 \end{cases}$$

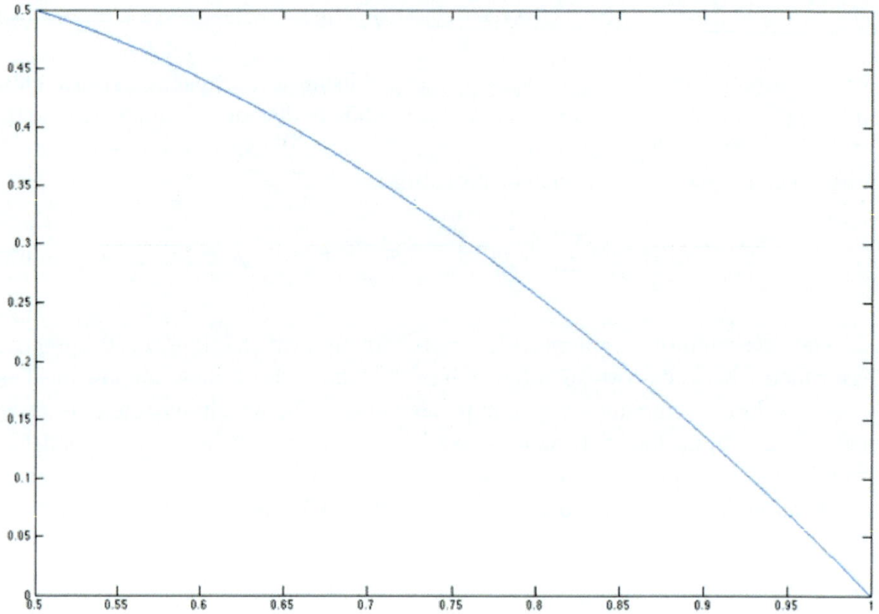

Fig. 3.2 The value of shannon entropy with the change of probability value

When a voxel is occupied certainly with $p_{m_r^i} = 1$, the entropy achieves minimum with 0. When the minimum value is achieved, it means the voxel is occupied certainly with no uncertainty.On the contrary, entropy reaches maximum value with 1 when the occupancy probability $p_{m_r^i} = 0.5$. When $p_{m_r^i} = 0.5$, it means it does not have any information of the voxel, the probability of the voxel is free or occupied is the same.

Then, the overall map uncertainty $H(m_r)$ is represented as the summation of probabilistic voxels:

$$H(m_r) = \frac{1}{n}\sum_{i=1}^{n} H(m_r^i) \tag{3.11}$$

3.2.2 *Uncertainty in Positional Value*

For registration, there is always a misalignment error between positional coordinates at each step. The misalignment error is hard to model due to the unknown amount of mismatch point between target scene and source scene. For the purpose of quantifying the error, it is assumed to be a Gaussian noise model.

As shown in Fig. 3.3, let d_{TS} be the true distance between target point m_r^i and source point n_S. At each iteration, point m_s^j iteratively attempts to find its ground truth point m_r^i, it would match to its nearest neighboring point m_r^c with a distance

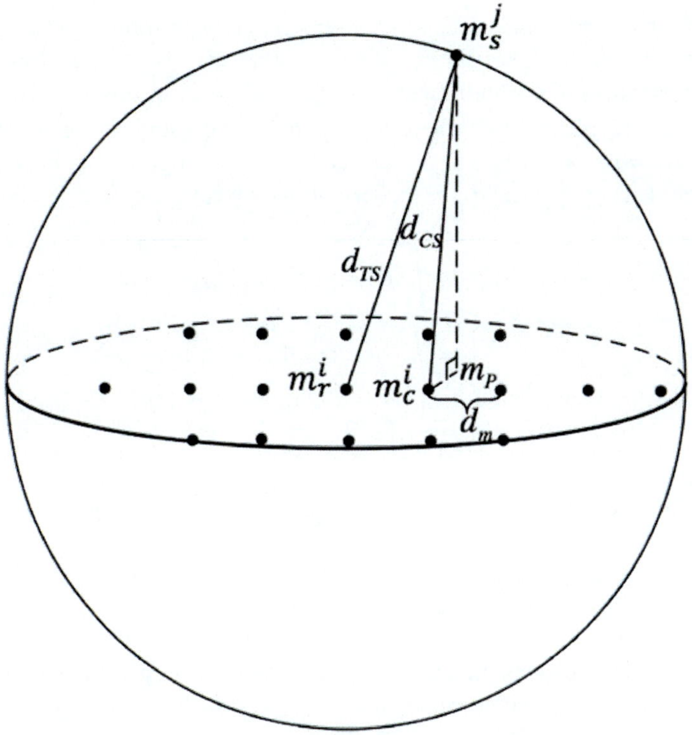

Fig. 3.3 Positional error model estimated in a spherical coordinate system

d_{CS}. So, there is always an offset between ground truth distance d_{TS} and the closet point distance d_{CS}.

In [1], conditional expectation is applied to model the error due to misalignment in each iteration. The author assumes that the points around the ground truth point m_r^i are continuously distributed. However, the points are discretized with equal distance d_{m_i} in our case.

To model the true distance d_{TS} between point m_r^i and point m_s^j using closet point distance d_{CS}, the registration error is assumed to be independent of noise and equal in three directions.

$$E(d_{TS}^2) = \sigma_{TS}^2 = \frac{1}{3}\sigma_{TSx}^2 = \frac{1}{3}\sigma_{TSy}^2 = \frac{1}{3}\sigma_{TSz}^2 \tag{3.12}$$

We assume the scene point m_s^j is uniformly distributed on the surface of a sphere with the radius of d_{TS}, so the probability density function is denoted as:

$$f(d_{CS}^2|d_{TS}) = \frac{1}{4\pi d_{TS}^2} \tag{3.13}$$

In 3.14, the conditional expectation of d_{CS}^2 given d_{TS} is formulated and calculated by integrating in the spherical coordinate system. d_{CS}^2 is computed as the length of the hypotenuse of a right triangle $\Delta_{n_s m_c m_P}$ as shown in Fig. 3.3. The true positional distance(unknown) d_{TS} between m_r^i and m_s^j is estimated using closet point distance(known) d_{CS}, where d_{CS} is computed as the length of the hypotenuse of a right triangle $\Delta_{m_s^j m_c^i m_P}$ and d_{m_i} is the size of the voxel.

$$\begin{aligned} E(d_{CS}^2|d_{TS}) &= \int_S d_{CS}^2 \cdot f(d_{CS}^2|d_{TS})ds \\ &= \int_S ((d_{TS}\cos\Phi)^2 + d_E^2)\cdot\frac{1}{4\pi d_{TS}^2}ds \\ &\approx \frac{1}{3}d_{TS}^2 \end{aligned} \tag{3.14}$$

In conditional expectation, $E[E(X|Y)] = E(X)$, so the expectation of d_{TS}^2 can be calculated as:

$$E(d_{TS}^2) \approx E(3E(d_{CS}^2|d_{TS}) - 2d_m^2) = 3d_{CS}^2 \tag{3.15}$$

The term $-2d_m^2$ indicates the error caused by discretized point cloud. Since mean square error(MSE) is the global estimate of d_{CS}, $d_{CS} \approx MSE$. As each dimension is identical, $\sigma_{TSx}^2 \approx MSE^2$.

Hence, we get the final form for the weight as:

$$\omega_o = k \cdot \frac{\sigma_{TSx}}{H} \approx k \cdot \frac{MSE}{H_{m_r}} \tag{3.16}$$

where k is positive constant factor. As the registration iteratively minimizes the error function, MSE should be non-increasing. Then weight ω_o is non-increasing with invariant H, and should tend to zero as the OICP converges to a perfect match. When the positional error is large, more weight is given to features. As the two maps get closer, the algorithm tends to rely more on positional value.

3.3 Transformation Evaluation and Probability Fusion

The transformation matrix T transforms source map m_s from local coordinate frame into an target map m_r coordinate system. Therefore, the voxel $m_s^j \in m_s$ is transformed to the coordinate frame of m_r by the transformation matrix T, which is

$${}^r m_s^j = T \cdot m_s^j \tag{3.17}$$

In theory, the voxel ${}^r m_s^j$ should coincide with its closet voxel m_r^i in map m_r. Due to the sensor noise in map and the misalignment error caused by OICP, an error between the corresponding pairs is always expected. Hence, a transformation evaluation criteria is proposed to measure the misalignment and further fuse the map with a relative entropy filter.

3.3.1 Transformation Evaluation

Although the registration algorithm will achieve convergence, there is no guarantee that a global minimum will be found and there is a high chance that the algorithms would fall into a local minimum.

In [2], an environment measurement model (EMM) is proposed to verify the transformation by calculating the Mahalanobis distance between matched points. Then the inliers percentage is set as a threshold to decide whether to accept or reject the transformation. For the statistic test, the null hypothesis assumption is that after performing the transformation, the individual maps should be aligned correctly. Let's consider the matching between voxels under mahanalobis distance. Then Mahalanobis distance between point m_r^i and ${}^r m_s^j$ can be written as:

$$f(m_r^i, {}^r m_s^j) = f(m_r^i; {}^r m_s^j, \Sigma_{ij}) \tag{3.18}$$

When operating on the Octomap representation, the sensor noise has already been incorporated into the occupancy probability. Hence, we utilize the inverse of occupancy probability to represent the spherical covariance of a voxel.

Since the covariance is equal in three directions, rotation matrix has no impact on Σ_i and Σ_j

$$\Sigma_{ij} = R^T \Sigma_j R + \Sigma_i = \frac{1}{p_{m_r^i}} + \frac{1}{p_{{}^r m_s^j}} \tag{3.19}$$

Then for each correlated pairs, Mahalanobis distance can be computed as follows, where d is a normalizing factor:

$$f_{m_{ij}} = -\frac{({}^r m_s^j - m_r^i)^T \cdot ({}^r m_s^j - m_r^i)}{d \cdot (p_{m_r^i}{}^{-1} + p_{{}^r m_s^j}{}^{-1})} \tag{3.20}$$

Note that the covariance defined here is only a scale factor and reveals that Euclidence distance is a special case of Mahalanobis distance. That covariance equals to identity matrix, when $p_{m_r^i} = p_{{}^r m_s^j} = 1$. Then Eq. (3.20) becomes

$$f = ||{}^r m_s^j - m_r^i||^2 \tag{3.21}$$

The inliers I are decided by setting a threshold of Mahalanobis distance, and the rest are counted as outliers O. Then a confidence level is applied as a criteria for rejecting the wrong estimates by computing the overall fraction of inliers in Eq. (3.22).

$$q = \frac{I}{I + O} \tag{3.22}$$

It is shown in the experiments that the applying of environment measurement model(EMM) effectively rejects high erroneous estimates that would largely undermine the quality of the fused map.

3.3.2 Relative Entropy Filter

Under sensor noise and motion uncertainty, the same object may have different probabilities in separated maps. This is called duplicate landmarks in feature map. However, in voxel map, the features and objects have been discretized into many voxels. It's hard count how many voxels consisting of an object. However, itâĂŹs vital to establish a strategy to consider the dissimilarities and fuse the voxels into the global probability map.

To merge pair-wise voxels, the separation level between the voxels has to be measured and evaluated. The test is based on Kullback-Leibler (KL) divergence, which is a commonly used relative entropy filter to measure the difference between two probability. Here the occupancy probability $p_{m_r^i}$ and $p_{^r m_s^j}$ are modeled as discrete random variables (RVs), and the difference between the two RVs are computed as:

$$KL(p_{m_r^i} || p_{^r m_s^j}) = p_{m_r^i} \log \frac{p_{m_r^i}}{p_{^r m_s^j}} + (1 - p_{m_r^i}) \log \frac{1 - p_{m_r^i}}{1 - p_{^r m_s^j}} \tag{3.23}$$

$KL(p_{m_r^i} || p_{^r m_s^j})$ represents the difference between the two probabilities. If KL is within a certain threshold, the dissimilarity are modeled as random errors and still fuse the corresponding pairs by taking the average

$$p'^{i}_{fuse} = \frac{p_{m_r^i} + p_{^r m_s^j}}{2} \tag{3.24}$$

To update the map, the corresponding pairs are deleted and replaced by the newly fused voxel.

If KL is beyond the threshold with $p_{m_r^i} > p_{^r m_s^j}$, it means the voxel with lower probability $p_{^r m_s^j}$ doesn't have enough information for the environment and will be rejected, then we only trust the voxel with higher probability and let $p'^{i}_{fuse} = p_{m_r^i}$.

The overall algorithm for transformation evaluation and probability fusion is summarized as follows:

Algorithm 1 Transformation Evaluation and Probability Fusion

Require: transformation matrix T, map m_r and m_s
Ensure: the local maps can be fused into a global map
1. $\Sigma_{ij} = R^T \Sigma_j R + \Sigma_i$
2. Calculate mahanalobis distance $f_{m_{ij}}$
3. Calculate inlier percentage q
if $q > q_{thres}$ **then**
transformation is accepted, **continue**
if $KL(p_{m_r^i} || p_{r_{m_s^j}}) < KL_{thres}$ **then**
$p^i_{fuse} = \frac{p_{m_r^i} + p_{r_{m_s^j}}}{2}$
else if $p_{m_r^i} > p_{r_{m_s^j}}$ **then**
$p^i_{fuse} = p_{m_r^i}$
else
$p^i_{fuse} = p_{r_{m_s^j}}$
end if

(a) Pioneer robot (b) Partial map generated from WKW Lobby

Fig. 3.4 Pioneer robot and partial map generated from WKW Lobby

3.4 Experimental Results

Experiments conducted using both simulated and real data are presented in this section. For simulation, a robot equipped with a simulated 3D depth sensor traversed the environment twice to create two maps in Gazebo simulator using RTAB [3]. A Pioneer 3 AT robot was tele-operated twice in the WKW lobby at Nanyang Technological University to build two maps. The robot was equipped with a Hokuyo Laser range finder for pose estimation using Gmapping [4] and an Asus Xtion PRO for 3D perceptions, as shown in Fig. 3.4. The individually generated Octomaps were post-processed using C++ and Matlab to generate the fused maps. The size of Octomaps in both experiments was set to be 0.1 m. A summary of the setting up can be shown in Table 3.1.

Table 3.1 The summary of setting up of the experiments

Type	Environment	SLAM Method	Sensor	Voxel Size
Simulation	Gazebo simulator	RTAB	Asus Xtion	0.1 m
Real world Experiment	WKW lobby	GMAPPING	Asus Xtion	0.1 m

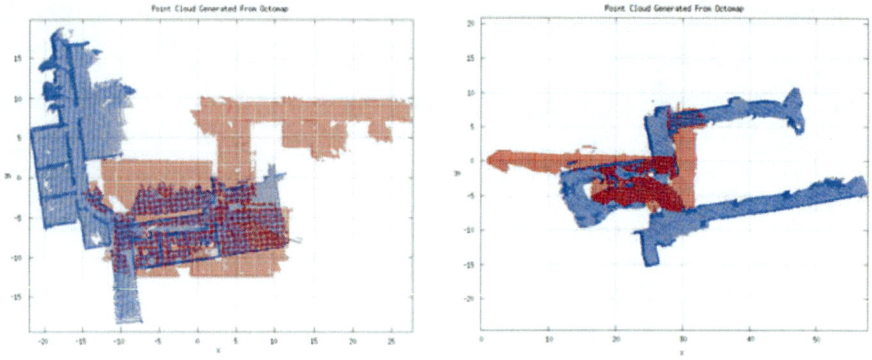

(a) Simulation (b) WKW Lobby

Fig. 3.5 Initial transformation for point cloud generated from Octomap

As mentioned before, the focus of the this chapter is on the fusion stage with an initial relative transform estimation. To simulate the availability of noisy initial relative transformations, two maps were generated with different initial transformations both in position and rotation. Figure 3.5 illustrates the point clouds extracted from the generated individual Octomaps with an en erroneous initial transformation. The dimension was about $50\,\text{m} * 30\,\text{m} * 10\,\text{m}$ for both maps.

3.4.1 Registration Results

In this section, the performance of OICP compared to ICP, point-to-plane ICP and normal distribution transform(NDT) is presented. As a baseline for comparison, standard ICP, point-to-plane ICP and NDT implementations from the Point Cloud Library (PCL) were used [5] and parameters were set according to the environments. The number of iterations for ICP, point-to-plane ICP and NDT is set to be 50. As mentioned before, the point clouds extracted from Octomap has a smoother local planes compared to raw sensor data. For OICP, point-to-plane ICP was performed for 15 iterations as initial registration stage to make use of this surface information, then ICP with occupancy probability formulated in this chapter was performed 35 iterations to get the point-wise correspondence.

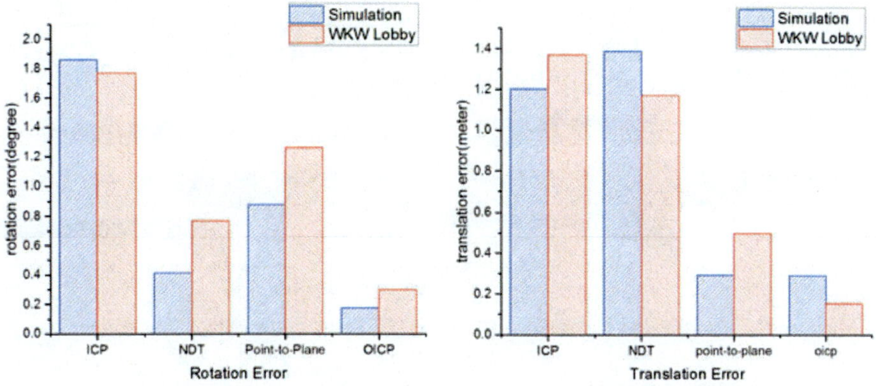

(a) Rotation Error (b) Translation Error

Fig. 3.6 Mean Error in both rotation and translation for the experiments, blue color represents the results conducted in simulation, while the results conducted in WKW lobby is shown in red

The comparison of accuracy of registration algorithms is summarized in Fig. 3.6. The translation error and rotation error are the Euclidean norm of and Euler norm of the difference between the ground-truth and the output.

OICP outperforms NDT, point-to-plane ICP, as well as map registration algorithm ICP used in [6]. Although point-to-plane ICP unable to provide point-wise correspondence, it makes use of the smooth plane and achieves better accuracy compared to default ICP based fusion. However, point-to-plane ICP still has drifts in translation and rotation. OICP combines surface and probability information to compute the point-wise correspondence with similar probability to generate the most accurate relative transformation. We should make special notice to NDT, which produces large errors, especially in translation. The reason might be the Gaussian distribution extracted from discretized sparse point cloud can't represent the environment explicitly compared to rich sensor data.

As described before, point-to-plane ICP and NDT can't provide point-wise correspondence. Hence only ICP and OICP are discussed in the following experiments. The registered results are shown in Figs. 3.7 and 3.8. A large offset in translation is shown for the result of ICP, and OICP produces approximately a perfect alignment. For both results, rotation errors are considered to be relatively accurate.

3.4.2 *Transformation Evaluation*

After computing a transformation, the accuracy is evaluated by calculating the percentage of inliers. Since ICP is deterministic with only coordinates, the covariance of each point is set to be identity matrix $I_{3\times 3}$. For map fusion, a strict criteria is set to guarantee the accuracy: the inliers threshold is set as 0.15 Mahalanobis distance and

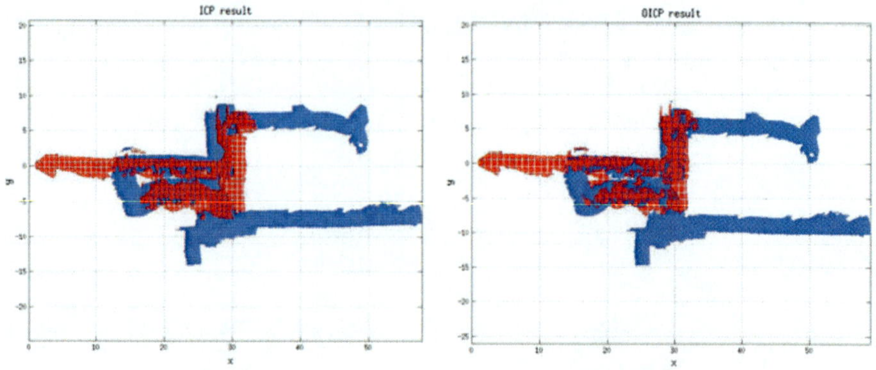

(a) ICP result of WKW Lobby (b) OICP result of WKW Lobby

Fig. 3.7 Registration result of WKW Lobby

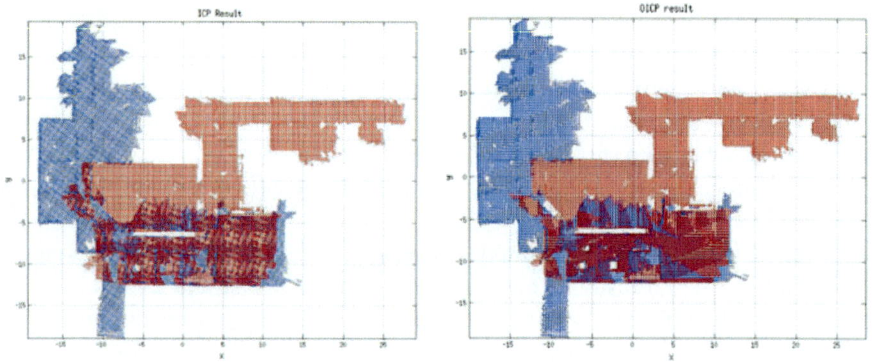

(a) ICP result of simulation (b) OICP result of simulation

Fig. 3.8 Registration result of simulated environment

the fraction of rejecting a transformation is set to be below 60%. Figure 3.9 illustrates the comparison of the average Mahalanobis distances between OICP and ICP in the two experiment settings. The box plot shows a box that encloses the middle 50% of the data, with the median value marked by a line and the mean value marked by a small square.

The inlier fraction based on the threshold is shown in Fig. 3.10. The inlier percentage of ICP is lower than OICP because ICP didn't converge to perfect alignment. Based on the selected threshold, OICP transformation generates an accurate and reliable map registration result compared to pure ICP based map registration approach in [6].

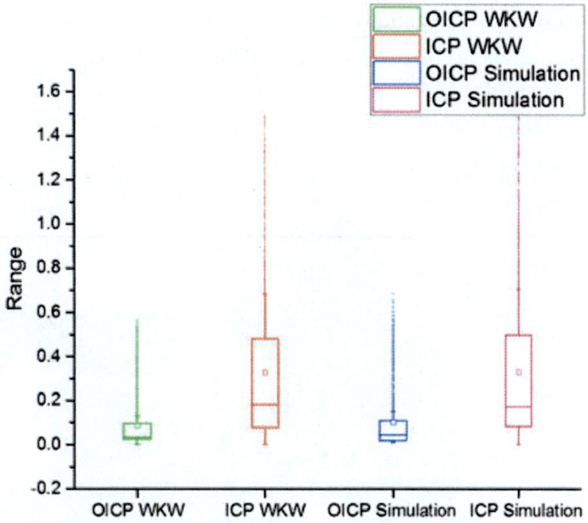

Fig. 3.9 Average Mahalanobis Distance

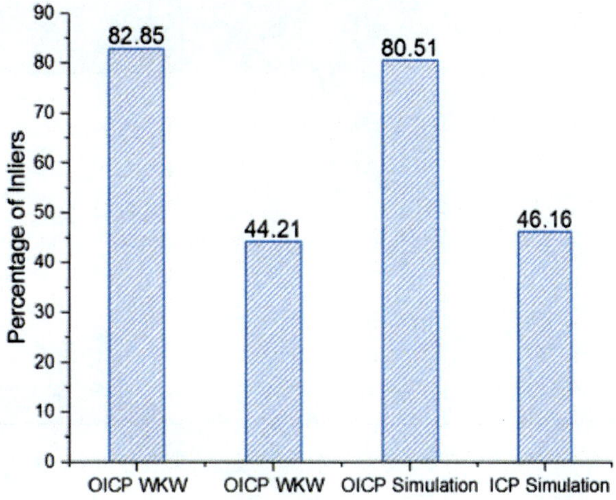

Fig. 3.10 The percentage of inliers

3.4.3 Probabilistic Map Fusion

After determining the correct transformation for map registration, a relative entropy filter is used to achieve map fusion by considering the dissimilarities between matched voxels. The relative entropy threshold is set as 0.1 in KL-divergence. In comparison, the fusion strategy in [6] is performed by directly taking the average of probability for all the matched voxels. In simulated environment, the entropy for map fusion by directly taking the average is 0.16. And the entropy of using relative entropy filter is 0.13. For real experiment in WKW Lobby environment, the entropy of directly taking the average is 0.24, and is 0.2 by using relative entropy filter. The results are summarized in Table 3.2.

Table 3.2 The entropy after map fusion

	Taking The Average	Relative Entropy Filter
Simulation	0.16	0.13
WKW Lobby	0.24	0.2

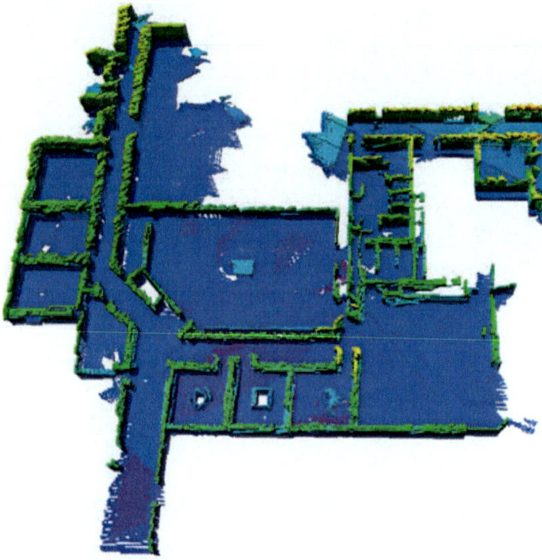

Fig. 3.11 The fused map of simulated environment

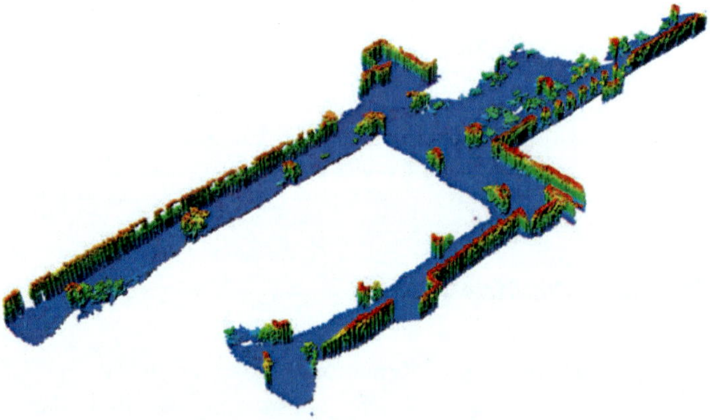

Fig. 3.12 The fused map of WKW Lobby

The entropy of WKW Lobby is higher than simulation due to real world noise and uncertainty. The experiment indicates that our map fusion strategy is able to combine probabilities of both maps effectively to further decrease the total uncertainty. Figures 3.11 and 3.12 show the final results of the fused 3D occupancy grid maps of simulation and WKW Lobby.

3.5 Conclusions

Most of the existing map matching algorithms ignore the probability information in the map. In this chapter, a probabilistic 3D map fusion algorithm Occupancy iterative closet point (OICP) was proposed. The OICP algorithm combines occupancy probability in conjunction with surface information to register point cloud extracted from Octomap. The weight between positional error and map uncertainty is controlled by modeling errors using Shannon entropy and positional misalignment. EMM is applied to evaluate the quality of transformation by rejecting low inlier fraction. The final map fusion is then achieved with a relative entropy filter which integrates the measurements of both maps and decreases the uncertainty of the global map. Experiments on simulated and real world environments were conducted and the evaluations confirmed that the proposed OCIP method is able to produce more consistent global 3D occupancy grid maps compared to existing methods in the literatures.

This chapter is based on the assumption that the local maps generated by single robot is consistent. However, this assumption may not hold true in real applications. In addition, OICP ignores the structure information in the map and only relies on the point cloud extracted. In Chap. 4, a two-level probabilistic map matching algorithm is proposed to combine structure edge information and individual voxel information.

References

1. Sharp GC, Lee SW, Wehe DK (2002) ICP registration using invariant features. IEEE Trans Pattern Anal Mach Intell 24(1):90–102
2. Endres F, Hess J, Sturm J, Cremers D, Burgard W (2013) 3-D mapping with an RGB-D camera. IEEE Trans Robot 30(1):177–187
3. Labbe M, Michaud F (2014) Online global loop closure detection for large-scale multi-session graph-based slam. In: 2014 IEEE/RSJ international conference on intelligent robots and systems, IEEE, pp 2661–2666
4. Grisetti G, Stachniss C, Burgard W (2007) Improved techniques for grid mapping with Rao-Blackwellized particle filters. IEEE Trans Robot 23(1):34–46
5. Rusu RB, Cousins S (2011) 3d is here: point cloud library (PCL). In: 2011 IEEE international conference on robotics and automation, IEEE, pp 1–4
6. Jessup J, Givigi SN, Beaulieu A (2015) Robust and efficient multirobot 3-d mapping merging with octree-based occupancy grids. IEEE Syst J 11(3):1723–1732

Chapter 4
Hierarchical Map Fusion Framework with Homogeneous Sensors

Fusing 3-D maps generated by multiple robots in real/semi-real time distributed mapping systems are addressed in this paper. A 3-D occupancy grid-based approach for mapping is utilized to satisfy the real/semi-real time and distributed operating constraints. This paper proposes a novel hierarchical probabilistic fusion framework, which consists of uncertainty modeling, map matching, transformation evaluation, and map merging. Before the fusion of maps, the map features and their uncertainties are explicitly modeled and integrated. For map matching, a two-level probabilistic map matching (PMM) algorithm is developed to include high-level structural and low-level voxel features. In the PMM, the structural uncertainty is first used to generate a coarse matching between the maps and its result is then used to improve the voxel level map matching, resulting in a more efficient and accurate matching between maps with a larger convergence basin. The relative transformation output from PMM algorithm is then evaluated based on the Mahalanobis distance, and the relative entropy filter is used subsequently to integrate the map dissimilarities more accurately, completing the map fusion process. The proposed approach is evaluated using map data collected from both simulated and real environments, and the results validate the accuracy, efficiency, and the support for larger convergence basin of the proposed 3-D occupancy map fusion framework.

4.1 Introduction

In Chap. 3, 3D volumetric map fusion algorithm occupancy iterative closet point (OICP) is proposed by extending the well known iterative closest point (ICP) algorithm. Algorithmically, it belongs to the scope of iterative dense registration algorithm, which is designed for raw sensor data registration. Due to the intrinsic nature

Y. Yue and D. Wang, *Collaborative Perception, Localization and Mapping for Autonomous Systems*, Springer Tracts in Autonomous Systems 2,
https://doi.org/10.1007/978-981-15-8860-0_4

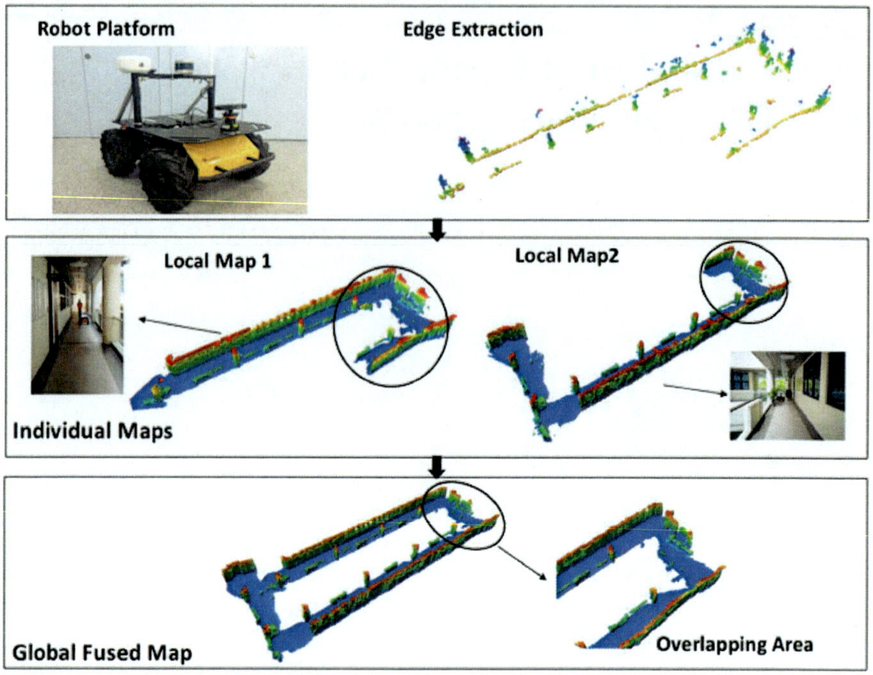

Fig. 4.1 An example of the proposed hierarchical probabilistic fusion framework

difference between raw sensor data and 3D occupancy grid maps, the direct use of dense registration algorithms in 3D occupancy grid map matching results in sub-optimal voxel associations thus sub-optimal matching. Since the research is based on the probabilistic map level, this chapter provides a comprehensive analysis of the map uncertainty modeling on high level geometric features and low level voxels. An example of the proposed framework is shown in Fig. 4.1. Top: Husky A200 robot platform equipped with Velodyne VLP-16 and the 3D edges extracted from the map. Middle: two individual maps of the WKW environment with the common area (black circle). Bottom: the fused global map and the zoomed in image of the overlapping area.

Generally, the *multi-robot 3D occupancy grid map fusion* consists of estimating the relative transformation between the local maps (i.e., map matching) and merging the occupancy map information (i.e., map merging), which are usually tackled independently of each other in this particular order. Existing 3D occupancy map matching approaches utilize some form of dense registration algorithm to estimate the relative transform between maps, where the algorithm proposed in Chap. 3 also falls within this scope. This is typically realized by converting the occupancy grid map into a point cloud, by representing each lowest level voxel in the 3D map by its center coordinate, and then employing a point set registration algorithm. The map merging step consists of verifying the accuracy of the estimated transformation and fusing the occupancy probabilities of the matching voxels. When a map matching algorithm returns an accurate transformation, the majority of map merging approaches merely resort to either stitching the overlapped voxels or to averaging the occupancy probabilities of matching voxels. While this two-step partitioning of 3D occupancy map fusion into a geometric problem and a probability fusion problem simplifies the solution and has seen significant development in recent years, it also introduces some issues that result in sub-optimal fusion results which need to be resolved. Besides, more detailed evaluation of these algorithms in the form of convergence and sensitivity to local map overlap changes warrants further examination and are the focus of this article.

Dense scan registration algorithms inspired initial solutions to map matching. These algorithms, such as ICP [1], NDT [2], GICP [3], OICP [4], are introduced to operate on raw scan data which are essentially samples of continuous surfaces with some noise. In the mapping process, sensor noise and robot localization errors are arbitrary digitized to generate 3D voxel maps with discontinuous and multiple voxel thick surfaces. Converting these voxel surfaces into the equivalent point clouds will not restore the continuous surfaces of samples as illustrated in Fig. 4.2. During the mapping process, motion uncertainty and sensor noise are accumulated and updated to a discontinuous occupancy grid map in the right image. Additionally, occupancy grid map formulation assumes voxel-wise independence to make the map update process computationally tangible. Ignoring the correlations among neighboring voxels with structural information also results in sub-optimal matching results.

To address the above challenges, the main objectives of the chapter are to develop a hierarchical probabilistic map fusion framework from a systematic viewpoint with the following contributions (see Fig. 4.3):

- The hierarchical probabilistic fusion system is proposed to address the integrated 3D occupancy map fusion problem, which is composed of three modules: uncertainty modeling, map matching and map merging.
- Based on the in-depth analysis and propagation of the 3D map uncertainty model, the two-level Probabilistic Map Matching (PMM) algorithm is developed by combining both high-level structural edge and low-level occupancy voxel features.
- The statistical testing model is applied to evaluate the matching result, where the acceptance threshold is auto-tuned based on the overlapping between the maps. Then, the individual maps are merged by combing occupancy probability based on a relative entropy filter.

The rest of the chapter is organized as follows: Sect. 4.3 then introduces our first contribution of modeling the uncertainty of 3D occupancy map data at voxel level and structural level as related to map fusion. The two-level probabilistic map matching framework based on the uncertainty modeling is detailed in Sect. 4.4. Section 4.5 presents the proposed statistical testing for transformation estimation and the relative entropy-based map merging approach. Section 4.6 presents the results of the experiments conducted using both simulated and real data. Section 4.7 concludes the chapter with a discussion on future work.

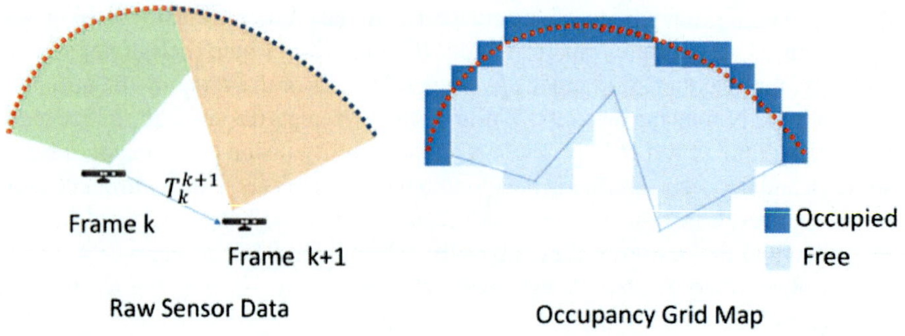

Fig. 4.2 The left image shows two consecutive scans of a continuous surface collected by the sensor

4.2 System Overview

The overview of the system architecture is depicted in Fig. 4.3. The focus of this chapter is to design the hierarchical system for map fusion, which can be divided into three modules, i.e., uncertainty modeling, map matching and map merging. The input of the system comes from the perception level, where a robot equipped with various sensors is utilized to map the environment. Prior to matching the maps, the individual voxel uncertainty and structural edge uncertainty are modeled and integrated to propagate the local uncertainty covariance of the 3D probabilistic map. Then, the map matching is performed with the hierarchical structure, where the matching starts from structural information layer and proceeds to full dense voxel map. The last step performs statistical testing to evaluate the quality of the transformation, where relative entropy filter probabilistic merging strategy is applied for successful estimations.

4.3 Map Uncertainty Modeling

This section exhibits the probabilistic mapping process and an efficient structural information extraction algorithm. Furthermore, the process of propagating and merging individual voxel uncertainty and structural edge uncertainty is presented.

4.3.1 Individual Voxel Uncertainty

Given an 3D occupancy voxel map G_r, and the information extracted from each voxel $(m^i_{rx}, m^i_{ry}, m^i_{rz}, L_{m^i_r}, d_{m_r})$ is defined as a tuple that includes the position of the extracted voxel center $(m^i_{rx}, m^i_{ry}, m^i_{rz})$, the occupancy probability value $p_{m^i_r}$ and the

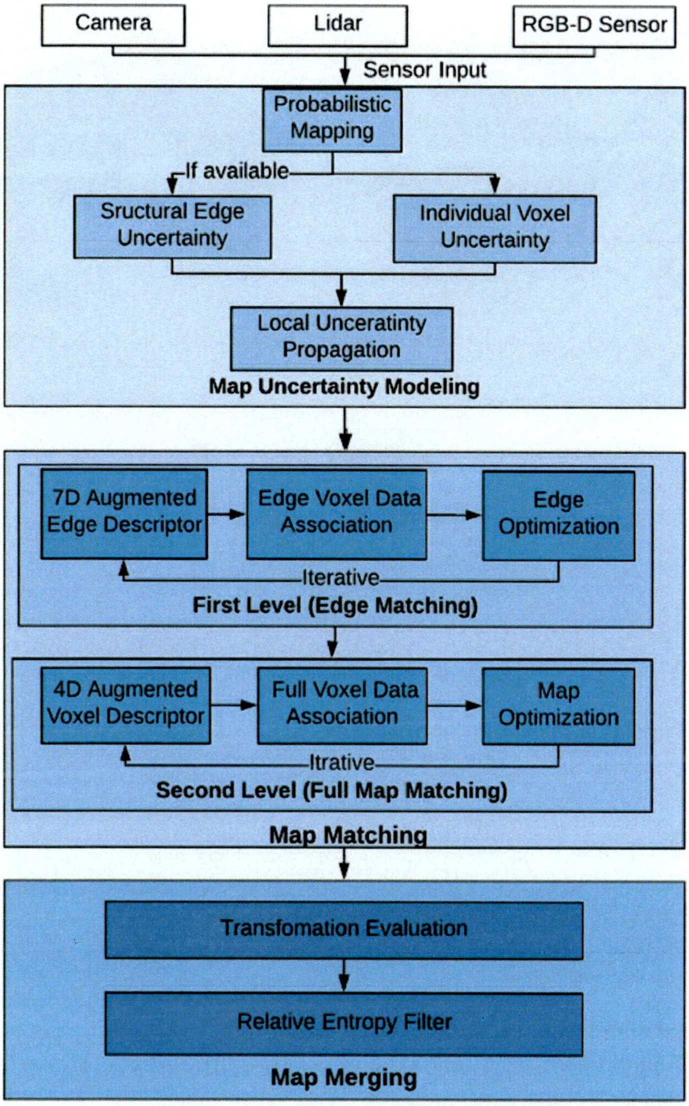

Fig. 4.3 The diagram of the system

voxel size d_{m_r}. Based on the independence assumption, the probability of a leaf node d_{m_r} given observation $z^i_{1:t}$ is updated by applying the Bayes filter (see Eq. (4.2)). The occupancy probability will update recursively when new sensor measurements are observed.

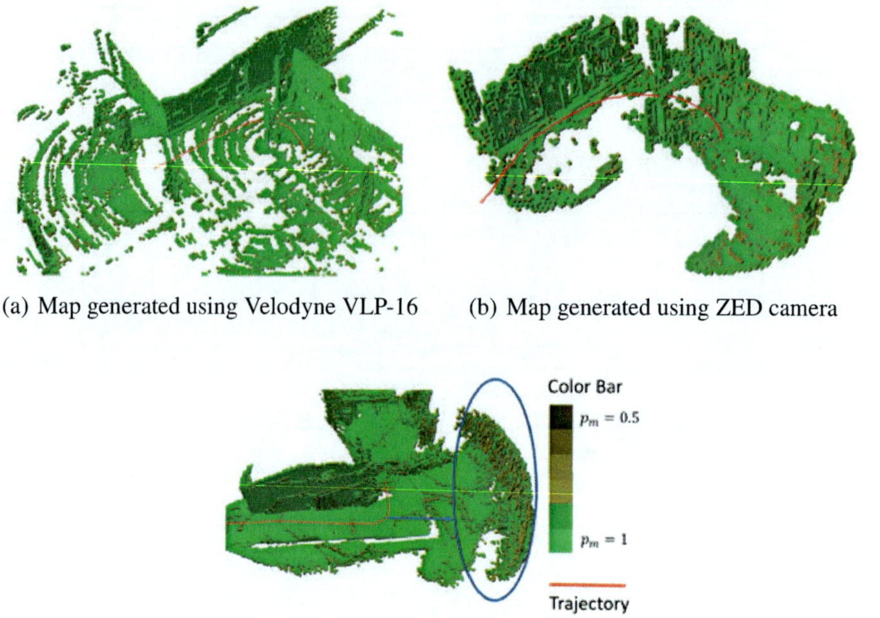

(a) Map generated using Velodyne VLP-16 (b) Map generated using ZED camera

(c) Map generated using ASUS RGB-D camera

Fig. 4.4 An example of the sensor uncertainty and motion uncertainty in the 3D map

$$p(m_r^i|z_{1:t}^i) = \left(1 + \frac{1 - p(m_r^i|z_t^i)}{p(m_r^i|z_t^i)} \frac{1 - p(m_r^i|z_{1:t-1}^i)}{p(m_r^i|z_{1:t-1}^i)} \frac{p(m_r^i)}{1 - p(m_r^i)}\right)^{-1} \tag{4.1}$$

$$\propto \underbrace{p(m_r^i|z_t^i)}_{\text{sensor model}} \cdot \underbrace{p(m_r^i|z_{1:t-1}^i)}_{\text{previous estimation}} \tag{4.2}$$

The updating of Eq. (4.2) depends on the current observation $p(m_r^i|z_t^i)$, the previous estimation $p(m_r^i|z_{1:t-1}^i)$ and the prior probability $p(m_r^i)$. The term $p(m_r^i|z_t^i)$ is the sensor model that represents the sensor uncertainty. Figure 4.4 illustrates an example of the sensor uncertainty and motion uncertainty, where the color bar represents the occupancy probability. Figure 4.4a and b show the map generated in the same environment with different sensors, which represents the uncertainty caused by the specification of the sensors. Since the raw sensor data registration only considers the sensor uncertainty, what makes probabilistic map matching different from sensor data dense matching is the term $p(m_r^i|z_{1:t-1}^i)$. The previous estimation term $p(m_r^i|z_{1:t-1}^i)$ models the accumulation of previous observations and motion uncertainty. As shown in Fig. 4.4c, large error with several thicks of walls is introduced when a sharp turned happens.

In probabilistic updating process, the only output is the occupancy probability value $p(m_r^i|z_{1:t}^i)$, where sensor readings and previous estimations are integrated into occupancy probability. To model the covariance of a voxel, the inverse of occupancy probability is utilized to represent the spherical covariance of a voxel. The Gaussian distribution $\mathcal{N}_{p_i}(\mu_{m_r^i}, \Sigma_{p_i})$ is located at the center of the voxel and identical in each direction, as shown in Fig. 4.6.

$$\Sigma_{p_i} = \begin{bmatrix} p_{m_r^i}^{-1} & 0 & 0 \\ 0 & p_{m_r^i}^{-1} & 0 \\ 0 & 0 & p_{m_r^i}^{-1} \end{bmatrix} \tag{4.3}$$

4.3.2 Structural Edge Uncertainty

The edge voxels in a 3D occupancy grid map are the voxels in the 3D map that exhibit a sudden large change in local curvature (i.e. sudden bend in the surface) and are extracted following the steps described in Alg. 2. First the surface normals of the occupied voxels are computed (lines 2–6). This is done by performing Principal Component Analysis (PCA) on the local neighborhood N_i of each occupied voxel m_r^i, defined by the radius r_N (line. 4). The output of this operation are the eigen values $\Lambda=\{\lambda_1, \lambda_2, \lambda_3\}$ in decreasing order and the corresponding principal component vectors $\mathbf{V} = \{\mathbf{v}_1, \mathbf{v}_2, \mathbf{v}_3\}$. The surface normal is $\mathbf{v}_3$ which is the principal component corresponding to the smallest covariance (i.e. hence smallest eigen value λ_3) (line. 5). Voxels with surface normals that on average has an angle cost larger than a given threshold κ to the surface normals within their neighboring voxels are selected as candidate edge voxels (line. 8). The average angle cost of a surface normal n_i with respect to its neighboring surface normals is computed using Eq. (4.4), where the angle cost between two normal vectors is computed using their cross product.

$$\overline{Cost_\theta(n_i)} = \frac{\sum_{m_r^j \in N_i} |\mathbf{n}_j \times \mathbf{n}_i|}{|N_i|} \tag{4.4}$$

A validity check is performed (lines. 9–20) in order to detect valid edge voxels from the candidate edge voxels. During this check, vectors are traced along the surface from each candidate voxel m_r^i in all directions of the local plane up to a distance d_T. Steps for initial trace vector generation for each m_r^i is given in lines. 11–13 where T_W^M is the transformation matrix from local frame W to global map frame M. t_1^M is the resultant vector after initial vector v_1^M is traced along the voxel surface from m_r^i up to d_T distance (line. 14). Value of 0.5m for d_T is used. If the resultant vectors t_1^M, t_2^M of any two opposing initial trace vectors v_1^M, v_2^M falls within the angle range $(\alpha_{min}, \alpha_{max})$, signifying a sudden bend in the local surface, m_r^i is detected as an edge voxel. Angle range is typically set to (80°, 100°). The detected voxels in E are statistically filtered (line. 23) to remove any remaining outliers to generate the edge voxels of the 3D map. Output of the edge extraction process is depicted in Fig. 4.5. The left image shows the input 3D occupancy grid map generated in an

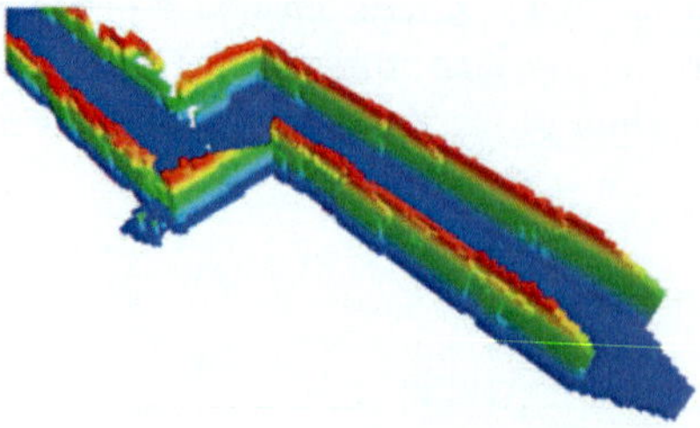

(a) The input 3D occupancy grid map

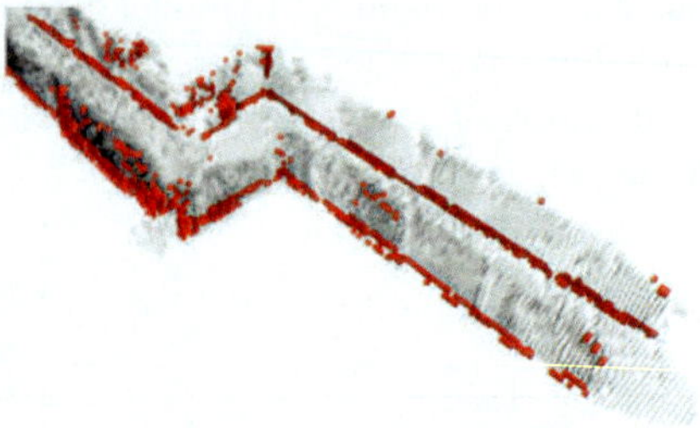

(b) The detected 3D edge voxels

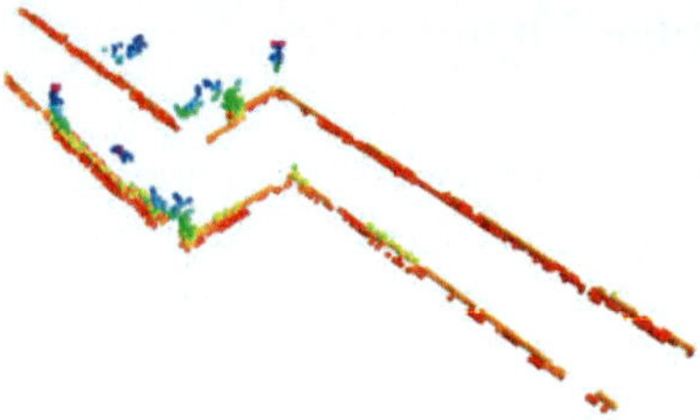

(c) Outlier removal and outputs 3D edge voxels

Fig. 4.5 The extraction of structural edge

indoor hallway, then the middle image presents the computed normals and detected 3D edge voxels from the 3D map. Finally, the outliers are removed and outputs the 3D edge voxels in the right image.

Given the detected edge voxels, local uncertainty Σ_{τ_i} of each edge voxel m_r^i is derived using the neighboring edge voxels. Then, the spatial Gaussian model $\mathcal{N}_{\tau_i}(\mu_{m_r^i}, \Sigma_{\tau_i})$ is located at the center of the edge voxel $\mu_{m_r^i}$. By performing eigen decomposition, covariance Σ_{τ_i} is factorized into $\Sigma_{\tau_i} = \mathbf{v_i}\lambda_i\mathbf{v_i}^T$. The eigen values $\lambda_i = \text{diag}(\lambda_i^1, \lambda_i^2, \lambda_i^3)$ in ascending order correspond to eigen vectors $\mathbf{v_i} = [\mathbf{v_i^1}, \mathbf{v_i^2}, \mathbf{v_i^3}]$, where the edge vector τ_i equals to $\mathbf{v_i^3}$. The edge covariance is shown in Fig. 4.6.

$$\Sigma_{\tau_i} = \left[\mathbf{v_i^1}\ \mathbf{v_i^2}\ \mathbf{v_i^3}\right] \cdot \begin{bmatrix} \lambda_i^1 & & \\ & \lambda_i^2 & \\ & & \lambda_i^3 \end{bmatrix} \cdot \left[\mathbf{v_i^1}\ \mathbf{v_i^2}\ \mathbf{v_i^3}\right]^T \tag{4.5}$$

4.3.3 *Local Uncertainty Propagation*

The significance of the structural information reconstruction comes at that, although the occupancy map is assumed to be voxel-wise independent, the continuous structural information is recovered by extracting the edges from the noisy map data.

Algorithm 2 Structural Edge Voxel Detection

Require: M^O : The occupied voxels of the 3D grid Map

$E = \phi$
for each $m_r^i \in M^O$ **do**
  $N_i = \{m_r^j : m_r^j \in M^O$ and distance$(m_r^i, m_r^j) < r_N \}$
  $(\Lambda, V) = \text{PCA}(\mathcal{N}_i)$
  $n_i = V[3]$
end for
for each $m_r^i \in M^O$ **do**
  if $\frac{\sum_{m_r^j \in \mathcal{N}_i} |\mathbf{n}_j \times \mathbf{n}_i|}{|\mathcal{N}_i|} > \kappa$ **then**
    create $W = (\mathbf{u}, \mathbf{v}, \mathbf{n}_i)$ right handed local coordinate frame at m_r^j
    for $\theta = \{0, \delta, \ldots, \pi - \delta, \pi\}$ **do**
      $\mathbf{v}_1^W = (cos(\theta), sin(\theta), 0, 1)^T$
      $\mathbf{v}_2^W = -v_1^W$
      $\mathbf{v}_1^M = T_W^M v_1^W$, $\mathbf{v}_2^M = T_W^M v_2^W$
      $\mathbf{t}_1^M$ = TraceVectorAlongSurfaceVoxels(m_r^i, v_1^M, d_T)
      $\mathbf{t}_2^M$ = TraceVectorAlongSurfaceVoxels(m_r^i, v_2^M, d_T)
      if $\alpha_{min} < acos(\mathbf{t}_1^M \cdot \mathbf{t}_2^M) < \alpha_{max}$ **then**
        $E = E \cup \{m_r^i\}$
        break
      end if
    end for
  end if
end for
$E = StatisticalOutlierRemoval(E)$
return E

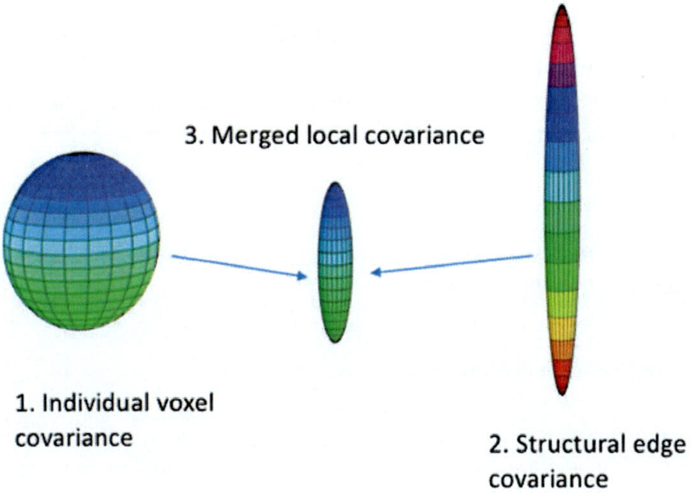

Fig. 4.6 The process of combing individual voxel covariance (left) and structural edge covariance (right), where the middle image shows the merged result (middle)

Hence, with the individual voxel covariance $\mathcal{N}_{p_i}(\mu_{m_r^i}, \Sigma_{p_i})$ and structural edge model $\mathcal{N}_{\tau_i}(\mu_{m_r^i}, \Sigma_{\tau_i})$ available, the two uncertainty models are integrated into an unified representation of the local uncertainty model $\mathcal{N}(\mu_{g_i}, \Sigma_{g_i})$:

$$\mathcal{N}(\mu_{g_i}, \Sigma_{g_i}) \sim \mathcal{N}(\mu_{m_r^i}, \Sigma_{p_i}) \cdot \mathcal{N}(\mu_{m_r^i}, \Sigma_{\tau_i}) \tag{4.6}$$

where Eq. (4.6) is the weighted averaging algorithm with the updated mean and covariance:

$$\mu_{g_i} = ({\Sigma_{p_i}}^{-1} + {\Sigma_{\tau_i}}^{-1})^{-1}({\Sigma_{p_i}}^{-1}\mu_{m_r^i} + {\Sigma_{\tau_i}}^{-1}\mu_{m_r^i}) = \mu_{m_r^i} \tag{4.7}$$

$$\Sigma_{g_i} = ({\Sigma_{p_i}}^{-1} + {\Sigma_{\tau_i}}^{-1})^{-1} \tag{4.8}$$

The process of integrating the uncertainty is shown in Fig. 4.6. The benefit of the uncertainty modeling and merging comes at two parts. Firstly, it accurately models the local uncertainty of the map with a properly designed Gaussian ellipse, which preserves the most valuable information in the environment. Secondly, the fusion combines the individual voxel covariance and structural edge covariance to represent the final uncertainty.

4.4 Two-Level Probabilistic Map Matching

To address 3D volumetric map registration, the two-level Probabilistic Map Matching (PMM) architecture is proposed. We start with formulating the two-level structure of the map matching algorithm, which starts from high-level edge features and moves to low-level full voxel matching. Then, map matching is further solved by decomposing it into establishing the probabilistic data association and optimizing the rigid transformation based on the data association.

4.4.1 The Formulation of Two-Level Probabilistic Map Matching Problem

4.4.1.1 Mathematical Definition

Problem *Given a set of 3D occupancy grid maps, each map can be denoted as* $G_r = \{m_r, g_r\}$, *where the point cloud* $m_r = \{m_r^i\}_1^N$ *and geometry features* $g_r = \{g_r^i\}_1^L$ *are extracted from the map* G_r. *The objective is to estimate the relative transformation* T *that maximizes the overlap between two maps.*

For dense registration of two maps G_r and G_s, the common solution is to fix one map G_r as the model map, and another map G_s as the scene map, where the rigid transformation T transforms the map G_s to the coordinate frame of G_r. The maximization of the overlapping between G_r and G_s is formulated as:

$$T = -\min_{T} \log p(G_r|T, G_s) \tag{4.9}$$

where $p(G_r|T, G_s)$ is the maximum likelihood estimation (MLE) that aims to find the most likely relative transformation T by matching the two partial maps G_r and G_s. The estimated matrix $T \in SE(3)$ is a rigid transformation matrix with rotation matrix $R \in SO(3)$ and translation vector $t \in \mathbb{R}^3$. The transformation T is iteratively minimized until a certain criteria is met.

4.4.1.2 Two-Level Map Matching Formulation

Here, the maps G_r and G_s are defined as 3D occupancy grid map and derive the structure for two-level matching. For the model map G_r, let full point cloud $m_r = \{m_r^i\}_{i=1}^{N_{m_r}}$ and geometry feature $g_r = \{g_r^i\}_{i=1}^{N_{g_r}}$. In this chapter, the geometry feature g_r^i refers to the coordinate of the voxel that belongs to the extracted edge. Instead of simply considering T as a fixed unknown parameter, it is modeled as a 6D random variable and estimated with maximum likelihood estimation (MLE). The registration of G_r and G_s formulated in Eq. (4.9) is computed in two levels, where edge matching and full map voxel matching are tackled sequentially in Eq. (4.10).

$$T = \min_{T} \left(-\underbrace{\log p(g_r|T, g_s)}_{\text{Level 1: Edge Matching}} - \underbrace{\log p(m_r|T, m_s)}_{\text{Level 2: Full Map Matching}} \right) \tag{4.10}$$

The two-level structure works sequentially. Firstly, in the first level, the extracted edges are matched by utilizing the weighted linear combination of positional distance, occupancy probability distance and edge direction distance into an overall distance metric. Then, the probabilistic quadratic function is solved with the nonlinear optimization algorithm. In the second level, full voxel maps are matched to achieve accurate transformation and derive the corresponding pairs for probability merging. Here, the chapter assumes voxel-wise correspondences to be independent, the probability density function in Eq. (4.10) can be simplified into Eq. (4.11), where $m_s^{c_i}$ and $g_s^{c_j}$ are the candidate corresponding voxels of m_r^i and g_s^i, respectively.

$$T = \min_{T} \left(\underbrace{-\sum_{j} \log p(g_r^j|T, g_s^{c_j})}_{\text{Level 1: Edge Matching}} \underbrace{-\sum_{i} \log p(m_r^i|T, m_s^{c_i})}_{\text{Level 2: Full Map Matching}} \right) \tag{4.11}$$

The problem of estimating the transformation matrix is often solved in two steps: data association and nonlinear optimization. Since the data association is unknown, the problem in Eq. (4.11) is iteratively estimated. First, the data association step efficiently estimates the hidden corresponding pairs by evaluating the similarity of the local descriptor. Then, the optimization step solves the rigid transformation based on the previous step by marginalizing the likelihood function.

4.4.2 Probabilistic Data Association

To estimate transformation T, data association between the model voxel m_r^i (g_r^j) and scene voxel $m_s^{c_i}$ ($g_r^{c_j}$) needs to be established, which is usually estimated by performing unidirectional nearest neighborhood search of Euclidean distance. With the presence of map noise, its hard to find the true correspondence with purely coordinate information. Therefore, geometrical features such as curvature, surface normal and point cloud density are introduced to find the accurate correspondence relationship.

In our previous work [4], occupancy probability is incorporated to establish voxel-to-voxel correspondences between all overlapped partial maps in $4D$ space. With the structural information extracted from Sect. 4.3, it will provide additional information to restore the corresponding relationship between two occupancy grid maps. Here, edge matching is augmented to $7D$ local descriptor in the first level with $\{m_i^x, m_i^y, m_i^z, p_{m_i}, \tau_i^x, \tau_i^y, \tau_i^z\}$, which includes the center coordinate of voxel ($3D$), occupancy probability ($1D$) and edge direction ($3D$).

The first level of the registration applies full 7D local descriptor with the distance metric defined in Eq. (4.12), the distance metric $d(g_r^i, g_s^j)$ is described by Euclidean point distance $d_e(g_r^i, g_s^j)$, occupancy probability distance $d_o(g_r^i, g_s^j)$ and edge geometry distance $d_\tau(g_r^i, g_s^j)$. The weights for the corresponding pair search are inversely proportional to the variance of the particular channel of the source information. The weighting vector $\omega = \{\omega_o, \omega_\tau\}$ is introduced to determine the importance of the occupancy probability and geometry distance relative to the positional distance.

$$
\begin{aligned}
d(g_r^i, g_s^j) &= d_e(m_r^i, m_s^j) + \omega_o d_o(p_{m_r^i}, p_{m_s^j}) + \omega_\tau d_\tau(\tau_r^i, \tau_s^j) \\
&= ||m_r^i - T \oplus m_s^j||^2 + \omega_o ||p_{m_r^i} - p_{m_s^j}||^2 + \omega_\tau ||\tau_r^i - T \oplus \tau_s^j||^2
\end{aligned} \tag{4.12}
$$

When it moves to the full map matching level, the structure information is not available. In this case, the dimension of descriptor reduced to $\{m_i^x, m_i^y, m_i^z, p_{m_i}\}$ in 4D space. The error metric defined in Eq. (4.13) denotes the error metric of full map matching, where $\omega_\tau = 0$.

$$
\begin{aligned}
d(m_r^i, m_s^j) &= d_e(m_r^i, m_s^j) + \omega_o d_o(p_{m_r^i}, p_{m_s^j}) \\
&= ||m_r^i - T \oplus m_s^j||^2 + \omega_o ||p_{m_r^i} - p_{m_s^j}||^2
\end{aligned} \tag{4.13}
$$

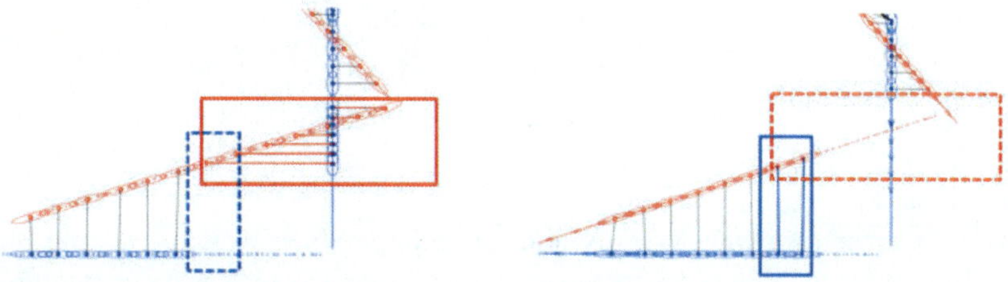

(a) Data association using Euclidean distance (b) Data association using 7D augmented distance

Fig. 4.7 An example of data association established between two L-shape datasets

The weighting vector $\omega = \{\omega_o, \omega_\tau\}$ measures the importance of the occupancy probability and geometry distance relative to the positional distance. The weight parameters $\{\omega_o, \omega_\tau\}$ are scale factors to determine the contribution of occupancy probability and geometry distance. Since weight ω_o has been decided in Chap. 3, the focus here is to estimate ω_τ. To model the two weights ω_τ, the uncertainties of structural information are estimated as described below.

4.4.2.1 The Uncertainty of Structural Information

For structural spatial covariance Σ_{τ_i}, the corresponding eigenvalues of $\lambda^i = \text{diag}[\lambda_1^i, \lambda_2^i, \lambda_3^i]$ is in ascending order. The eigenvalues denote the length of the principal coordinate axes, which are the structural uncertainties along three directions. More specifically, the largest eigenvalue λ_3^i corresponds to the edge vector τ_i. Here, Eq. (4.14) is defined to represent the uncertainty of the edge.

$$\lambda_\tau = \sum_i^n \frac{\lambda_3^i}{n(\lambda_1^i + \lambda_2^i + \lambda_3^i)} \tag{4.14}$$

When the edge is perfect without any noise on the perpendicular plane, $\lambda_\tau = 1$. And the extreme case is the three eigenvalues are equal, and the three axes have the same length, then $\lambda_\tau = \frac{1}{3}$. In the second level of full voxel map registration, $\lambda_\tau = 0$ as no structural information is available.

4.4.2.2 Overall Weights

With the estimated uncertainties of occupancy probability, structural information and positional error, the weights ω_o, ω_τ defined in Eq. (4.12), (4.13) are denoted as:

$$\omega_o = k_1 \cdot \frac{\sigma_{TSx}}{H_m} \approx k_1 \cdot \frac{MSE}{H_{m_r}} \tag{4.15}$$

$$\omega_\tau = k_2 \cdot \sigma_{TSx} \cdot \lambda_\tau \approx k_2 \cdot MSE \cdot \lambda_\tau \tag{4.16}$$

where k_1 and k_2 are the normalization factors. As the registration iteratively minimizes the error function, MSE is non-increasing. Since weights ω_o and ω_τ are constant in the matching process, the MSE should tend to zero as the matching converges to a perfect match. When the positional error is large, we should rely more on occupancy probability value and geometric information (only for the first level). As the two maps get closer, the algorithm tends to rely more on positional value.

To show the superiority of our 7D distance metric compared to purely Euclidean distance, an example of data association between two L-shape edges is presented in Fig. 4.7. In Fig. 4.7a, it only considers the Euclidean distance. While in Fig. 4.7b, the 7D augmented descriptor is applied. The full line red box shows the wrong data association estimated in Fig. 4.7a, which is removed in Fig. 4.7b. While the black box in the upper image shows the missed data association, which is found and connected with 7D descriptor in Fig. 4.7b. As can be seen, the data association using only Euclidean distance based closet neighborhood search falls into the false match, which might lead to wrong convergence in the optimization process. In contrast, the 7D distance metric combined with the augmented information filters out the wrong correspondences and finds correct pairs. The results prove that combining edge information, occupancy probability, and point coordinate tend to find accurate data association.

4.4.3 Error Metric Optimization

Based on the two-level structure, the registration strategy starts from the geometry structural edge layer and proceeds to the dense grid voxel layer. The optimization problem is to minimize the error function defined in Eq. (4.11) with the data associations estimated.

Starting with the edge optimization, robust correspondences have been established, and fewer voxels will be optimized. Applying the prolonged shape of the local uncertainty covariance Σ_{g_i} located along the edge direction will penalize two edges not lying in the same direction. The applying of local uncertainty covariance reduces the number of iterations compared to Euclidean distance based optimization algorithm. Then it forces the two edges to have the same direction with positional coordinate aligned, which enhances the convergence basin to inaccurate initial estimation and improves overall registration results.

$$T = \min \sum_j (g_r^j - T \oplus g_s^{c_j})^T (\Sigma_{g_r^j} + R^T \Sigma_{g_s^j} R)^{-1} (g_r^j - T \oplus g_s^{c_j}) \tag{4.17}$$

Based on the edge registration outputs, the dense voxel-wise correspondences will be established to refine the estimation. Here, the uncertainty model involved is the individual voxel covariance Σ_{p_i}. Since the covariance is equal in three directions, rotation matrix has no impact on Σ_{p_i}. Then the error function is simplified to the norm of Euclidean distance with the scale factor (see Eq. (4.18)). Since not all the corresponding error terms should have the same weight, larger weight is assigned to the associations with higher occupancy probability.

$$T = \min \sum_i ({p_{m_r^i}}^{-1} + {p_{m_s^{c_i}}}^{-1})^{-1} ||m_r^i - T \oplus m_s^{c_i}||^2 \tag{4.18}$$

4.5 Transformation Evaluation and Probability Merging

In this section, the statistical testing of accepting or rejecting the transformation is presented. Then, a relative entropy filter is proposed to measure the map dissimilarities and merge individual maps into a global map.

4.5.1 Transformation Evaluation

In Chap. 3 an environment measurement model (EMM) is applied to verify the transformation by calculating the Mahalanobis distance between matched points. Here, point cloud map m_s has been transformed to the coordinate frame of m_r and is denoted as ${}^r m_s$. Since the edge matching is a middle process, the evaluation is focused on the voxel-wise correspondences between full map voxels. The statistical evaluation of map m_r given ${}^r m_s$ is formulated in Eq. (4.19).

$$p(m_r|{}^r m_s) = \prod_i^{N_{m_r}} p(m_r^i|{}^r m_s^{c_i}) \tag{4.19}$$

A proper inconsistency distance should be defined to describe the discrepancy between submaps. As illustrated in Eq. (4.18), the probability of each correlated pair can be computed as follows:

$$p(m_r^i|{}^r m_s^{c_i}) = exp\left(-\frac{(m_r^i -^r m_s^{c_i})^T \cdot (m_r^i -^r m_s^{c_i})}{{p_{m_r^i}}^{-1} + {p_{m_s^{c_i}}}^{-1}}\right) \tag{4.20}$$

The inliers are decided by setting a threshold of Mahalanobis distance. Then, a confidence level is applied as the criteria for rejecting the wrong estimates by computing the overall fraction of inliers. The critical thing is to set a proper threshold for the ratio to accept the transformation. To address this issue, the parameter Θ is

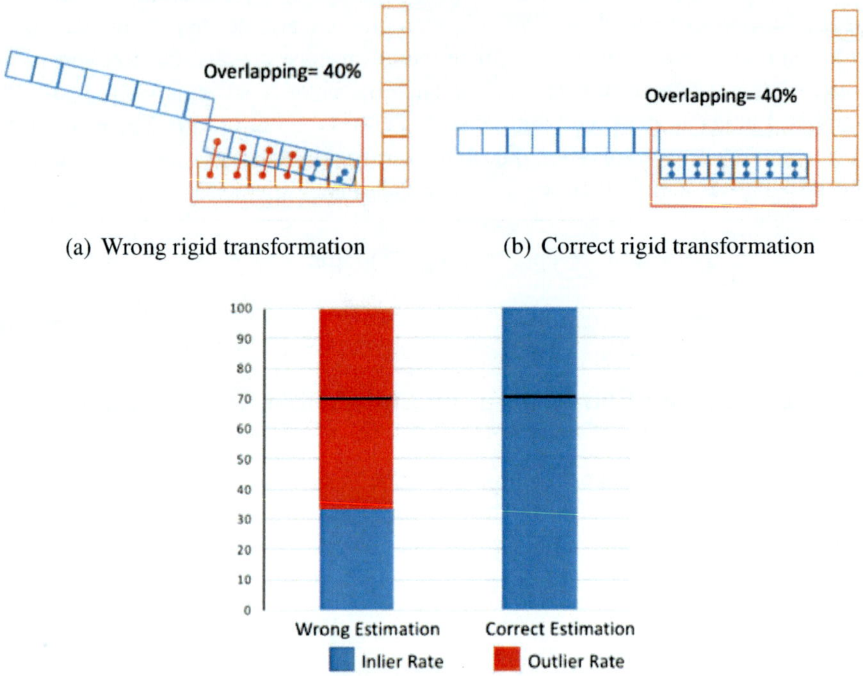

(a) Wrong rigid transformation

(b) Correct rigid transformation

(c) The plot of inlier and outlier rate, where the acceptance is 70%

Fig. 4.8 Given the same datasets with two different rigid transformations

introduced to measure the overlapping between the two registered maps. The value is calculated by dividing the number of matched pairs N_{match} by the number of voxels on two maps.

$$\Theta = \frac{1}{2} \cdot \left(\frac{N_{match}}{N_{m_s}} + \frac{N_{match}}{N_{m_r}} \right) \tag{4.21}$$

Since accepting a wrong rigid transformation will lead to a diverged global map, a conservative acceptance rate is set to be 60%. Besides, as the overlapping decreases, the acceptance rate is enhanced with special care for extreme low overlap cases. The relationship between the overlapping metric Θ and the acceptance threshold of the transformation is defined as follows:

- $\Theta < 20\%$, the inlier percentage should $>80\%$
- $20\% < \Theta < 60\%$, the inlier percentage should $>0.9-\frac{\Theta}{2}\%$
- $\Theta > 60\%$, the inlier percentage should $>60\%$

This simple relationship leads to satisfactory performance in our experiments. An example in Fig. 4.8 shows the applying of statistical testing efficiently rejects

Table 4.1 Summary of the threshold parameters

Notation	Location	Value
κ	Alg. 2	Set as 0.5
$d(g_r^i, g_s^j)$	Eq. (4.12)	Set as 2
$d(m_r^i, m_s^j)$	Eq. (4.13)	Set as 0.5
Inlier percentage	Eq. (4.20)	Depending on overlapping metric Θ
$KL(o_{m_r^i} \| o_{m_s^{ci}})$	Eq. (4.22)	Set as 0.1

high erroneous estimates that would largely undermine the quality of the fused map. Statistical testing is performed to decide whether to accept the transformation.

4.5.2 *Relative Entropy Filter*

We introduce $o_{m_r^i}$ as a discrete random variable (RV) with two states: occupied and free, where $p(o_{m_r^i} = \text{occupied}) = p_{m_r^i}$, $p(o_{m_r^i} = \text{free}) = 1 - p_{m_r^i}$, and the difference between the two RVs are computed as:

$$KL(o_{m_r^i}||o_{m_s^{ci}}) = p_{m_r^i} \log \frac{p_{m_r^i}}{p_{m_s^{ci}}} + (1 - p_{m_r^i}) \log \frac{(1 - p_{m_r^i})}{(1 - p_{m_s^{ci}})} \tag{4.22}$$

As mentioned in Chap. 3, Kullback-Leibler (KL) divergence is used relative entropy filter to measure the difference between two random variables. $KL(o_{m_r^i}||o_{m_s^{ci}})$ represents the difference between the two probabilities. If KL is within a certain threshold, the dissimilarity is seemed as random errors and the corresponding pairs are fused with weights

$$p^i_{fuse} = \frac{p_{m_r^i}}{p_{m_r^i} + p_{m_s^{ci}}} p_{m_r^i} + \frac{p_{m_s^{ci}}}{p_{m_r^i} + p_{m_s^{ci}}} p_{m_s^{ci}} \tag{4.23}$$

If KL is beyond the threshold with $p_{m_r^i} > p_{m_s^{ci}}$, it means the voxel with lower probability $p_{m_s^{ci}}$ does not have enough information for the environment and will be rejected, only voxel with higher probability is considered and let $p^i_{fuse} = p_{m_r^i}$. For parameter settings for the equations, please refer to (Table 4.1).

4.6 Experimental Results

Experiments conducted in simulated and real-world environments are summarized in this section. Both qualitative and quantitative results are presented in the following aspects:

(a) Evaluation of how the proposed approach improves the accuracy in both translation and rotation estimation compared with the baseline algorithms, where fewer iterations required to achieve the same accuracy.

(b) Exploration of the effect of reducing the overlap between the two maps showing that extracting the structural information increases convergence.

(c) Showing that the efficiency of the proposed algorithm is significantly improved compared to the baseline algorithms, proving that real-time performance is guaranteed.

(d) Presenting that our algorithm increases the robustness under initial perturbations, where statistic results show the improved performance on accuracy and convergence.

(e) Evaluation of how the statistical testing accepts the correct transformation and merges the map using relative entropy filter.

As mentioned before, the research is focused on the fusion stage with an initial relative transform estimation. The initial relative transformation between the coordinate frames of different robots can be extracted during the deployment phase of the mission if such information is available. Else it can be extracted later on through coordinated rendezvous actions.

To simulate the noisy initial relative transformations, two maps with different initial transformations both in translation and rotation were generated.

4.6.1 Evaluation Protocol

Experiment Environment: Experiments conducted using simulated and real data are compared by operating the robot to collect data and generate maps in three different scenarios in Nanyang Technological University (see Table 4.2). Each robot was equipped with a Hokuyo Laser range finder for pose estimation using Gmapping, Asus Xtion Pro RGB-D camera for 3D perception.In the simulation dataset Willow, the robot was tele-operated to explore environment twice from the same starting point. The Willow is a standard ROS/ Gazebo office world with varying room sizes and corridors. In the real environments Corridor and WKW, the recorded datasets were divided into two segments to generate two different maps with some overlap. The

Table 4.2 Summary of the experiment environment

Dataset	Dimension(x*y*z)	Map Size	Edge Size	Environment Type
Corridor Dataset	20m*5m*3m	46000 voxels	**3500** voxels	Structural, static
WKW Dataset	40m*15m*3m	67071 voxels	**4300** voxels	Semi-stru, dynamic
Willow Dataset	50m*40m*3m	76000 voxels	**3200** voxels	Structural, static

corridor is a long corridor in the building, while the WKW is the lobby of a building with walking people and clutter furniture/objects. The ground truth transformation between the maps was acquired by recording the experimental setup. The resolution of the 3D occupancy grid map was set to be 0.1m.

Error Metric: The chapter compares the estimated transformation T_e to the ground truth T_g. Then the error ΔT is calculated as $\Delta T = T_e \cdot T_g^{-1} = \{\Delta R, \Delta t\}$. The 3D translation error $e_t = ||\Delta t||$ and 3D rotation error $e_r = ||\Delta R|| = \arctan \Delta R$ are the Euclidean norm of and Euler norm of the difference between the ground-truth and the output. Another metric MSE (Mean Squared Error) is the mean squared Euclidean distance between the matched pairs.

Comparison Baseline: To demonstrate the superiority of the two-level PMM algorithm, the comparison is conducted on two levels. Firstly, the ICP based edge matching is compared against the first level of PMM. Then, the full PMM is compared against ICP, point-to-plane (P2P) ICP, Generalized ICP (GICP) and Normal Distribution Transform (NDT) based dense registration algorithm. The implementation of the compared algorithms are from the standard Point Cloud Library [5].

4.6.2 Edge Matching Analysis

One of the novelties of the proposed first level PMM comes at the extraction of structural information and modeling into a probabilistic registration problem. An alternative way to register extracted edges is to directly apply point-to-point ICP algorithm, which ignores the local uncertainty of each edge voxel. Here, the first level of PMM is compared against the ICP edge matching on two datasets. The initial perturbations were randomly generated for the two experiments, and the registration results of convergence rate and matching error are exhibited.

4.6.2.1 Registration Accuracy

The experiment was conducted in two datasets by extracting the edges from the full 3D map. Figure 4.9a shows the ICP edge matching trapped into local minimum after several iterations. The reason is due to the wrong data association established, as shown in the black line in the figure. Therefore, the algorithm fails to converge to the right global minimum. Compared with ICP edge matching, the first level of PMM makes full use of the constraint on x and y directions and combines the local uncertainty information. Hence, the correct correspondences are established and gradually converged to the desired position.

In the WKW dataset, both algorithms converge to the correct minimum (see Fig. 4.9b). However, the algorithm proposed requires fewer iterations compared with ICP edge matching. In the experiment, our algorithm converges to the local minimum after around 15 iterations, while the classical ICP requires more than 25 iterations. In summary, our algorithm increases the efficiency of convergence speed

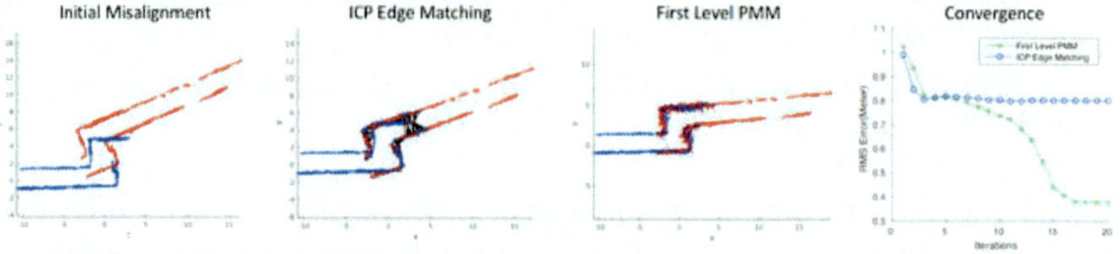

(a) Registration results of the edges extracted from Corridor dataset. The black lines show the correspondences established between the two edge maps.

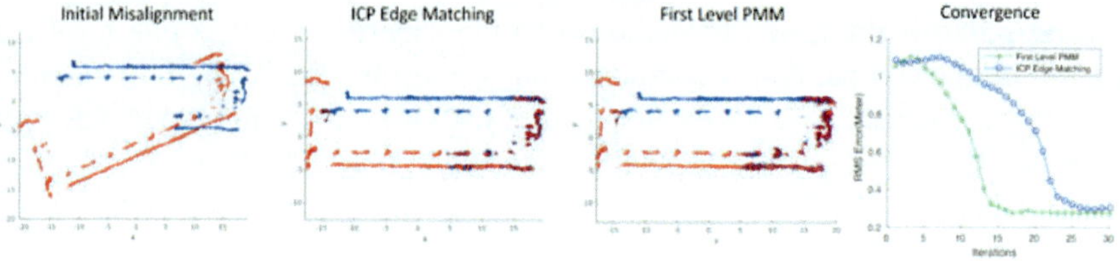

(b) Registration results of the edges extracted from WKW dataset. Both algorithms converge to the correct position, where the first level of PMM converges much faster.

Fig. 4.9 Comparison between edge registration algorithms, where first level of PMM is compared against ICP based edge registration algorithm

by nearly 40%. The reason is that the proposed optimization algorithm penalizes more on the offset between the edge direction and increases the convergence speed.

4.6.2.2 Sensitivity to Overlapping

Since the maps are viewpoints independent and the assumption of large overlapping in dense registration does not hold true here. The influence on different overlapping parameters is also tested. The original overlapping rate between the two maps in Fig. 4.9b is 45%. Then, fixing the scene map to remove part of the map data by decreasing 1m gradually on x-direction of the scene map for nine iterations ($x = -9$m). The removing of the map data leads to the decrease of the overlapping rate to 10% when $x = -9$m. In Fig. 4.10, the change of rotation and translation error with the dropping of the overlapping rate is summarized, where our algorithm shows a larger convergence basin. In contrast, the classical ICP based approach has a large sliding error in the x-direction. The result shows proposed algorithm can make the most use of the map information and increase the region of attraction of the correct minimization direction.

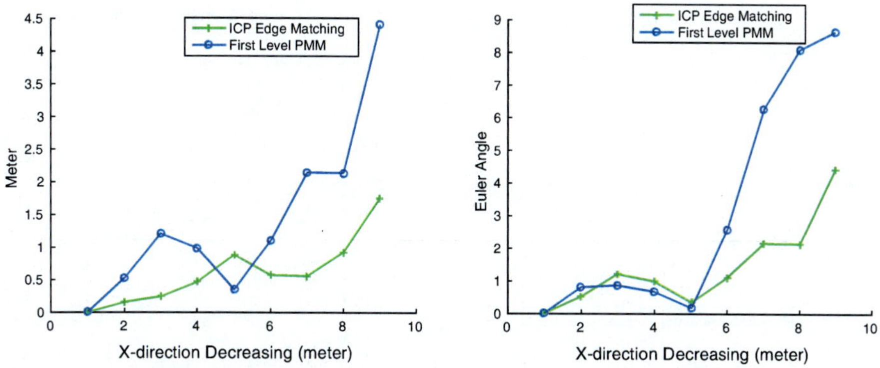

(a) Translation Error (b) Rotation Error

Fig. 4.10 Sensitivity to the change of overlap in WKW dataset, where ICP edge matching and first level PMM are compared

4.6.3 Full Map Matching Analysis

For the three experiments above, an exhaustive comparison between ICP map matching, P2P ICP map matching, GICP map matching, NDT map matching, first level PMM and full PMM has been carried out to test the results of registration accuracy and sensitivity to different initial inputs. The results of registration on three different scenarios are compared against the ground truth relative transformation. The scene map was transformed away from the ground truth and used as input to test the convergence. For each testing pairs, the offsets between -2m and 2m in steps of 1m (along x and y-axes) were introduced. In addition, the maps were rotated from -15° to 15° (in steps of 7.5°). Thus, a total of 125 test transformations were generated and applied to the maps. For all algorithms, the threshold for nearest neighbor search is 2m, and the number of iterations is set to be 30.

4.6.3.1 Sensitive to Initial Perturbations

The boxplot presented in Fig. 4.11 supports that the full PMM algorithm outperforms to those of ICP, P2P ICP, GICP, and NDT. The boxplot encloses the middle 50% of the data, with the median value marked by a line and the mean value marked by a green rhombus. The red cross marks indicate outliers, i.e., data values below one standard deviation above and below the mean of the data.

For the first scenario, the translation and rotation errors of the algorithms except full PMM present a significant variance, indicating that the optimization process often falls into the local minimum. The reason is that the corridor environment has low translational features in the y-direction, where other algorithms fail to utilize the y-direction constraint in the optimization process. In the second environment,

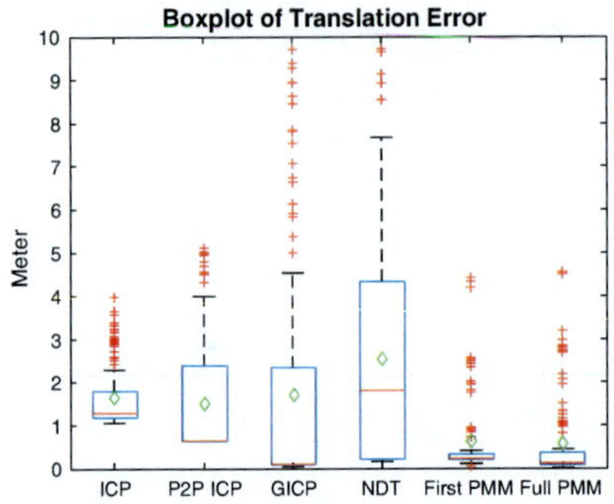

(a) Corridor dataset translation error

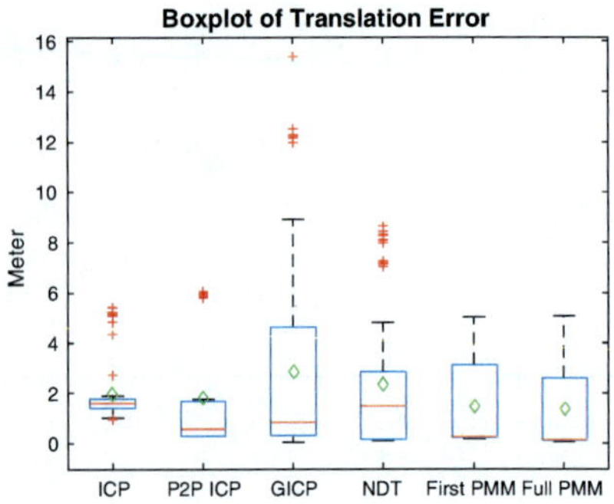

(b) WKW dataset translation error

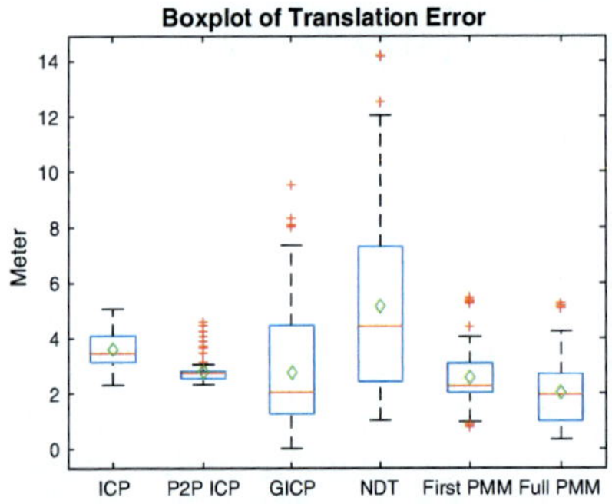

(c) Willow dataset translation error

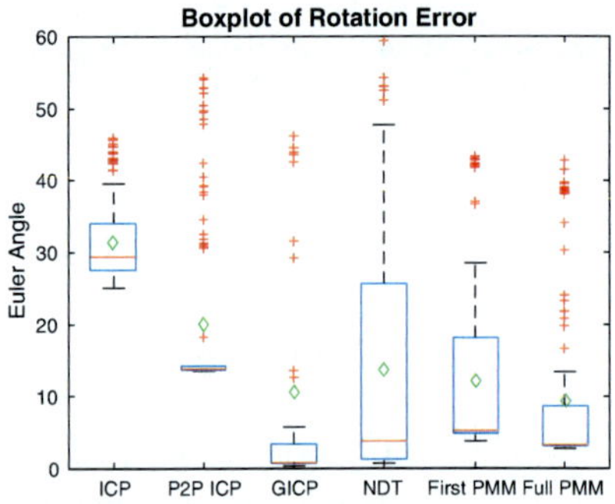

(d) Corridor dataset rotation error

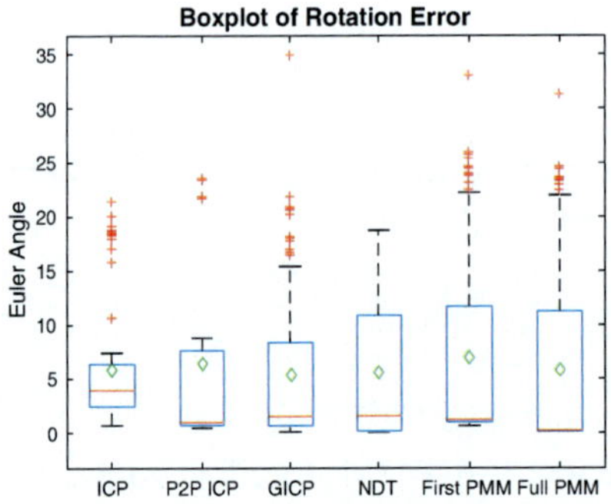

(e) WKW dataset rotation error

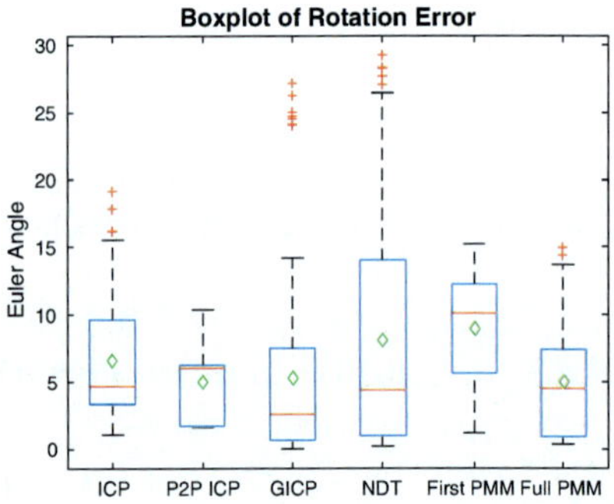

(f) Willow dataset rotation error

Fig. 4.11 Summary of convergence of the evaluated methods under 125 initial perturbations. Translation error (top) and rotation error (bottom)

Table 4.3 The mean error in both rotation and translation estimations. (The best performance is denoted in bold.)

Dataset	ICP		P2P ICP		GICP		NDT		First PMM		Full PMM	
	Euler	Meter	Euler	Meter	Euler	Meter	Euler	Meter	Euler	Meter	Euler	Meter
Corridor	31.39	1.63	19.97	1.51	10.58	1.71	13.63	2.56	12.06	0.62	**9.37**	**0.57**
WKW	5.77	1.95	6.31	1.78	5.32	2.84	**5.51**	2.32	6.90	1.47	5.60	**1.34**
Willow	6.54	3.60	**4.97**	2.78	5.28	2.72	8.07	5.13	8.88	2.54	4.99	**2.00**

all algorithms present acceptable performance. Thanks to the abundant translational features on x-direction and y-direction, all the algorithms can relatively converge to the correct transformation. In the third environment, large translation errors are found for all the algorithms. This is due to locally similar geometries, where the small office rooms are very similar. The rotation errors are much lower, suggesting that in the structured environment the translation errors are the primary concern.

4.6.3.2 Convergence Analysis

Table 4.3 summarizes the mean error in both rotation/translation estimations, the full PMM outperforms other algorithms in most of the cases. The exception is the rotation estimation of P2P ICP slightly exceeds full PMM in Willow dataset, and NDT slightly exceeds full PMM in WKW dataset. Besides the performance of full PMM, the GICP achieves the second best results. Since not all the outputs are successful, the criteria for a correct registration was chosen as a translation error of 0.6m or less and a rotation error of 6° or less.

The performance summarized in Fig. 4.12 exhibits the successful fraction of all three datasets. The trend is clear that the smallest errors in Table 4.3 exhibit the highest percentage of success. In Corridor dataset, shows poor performance on ICP and P2P ICP, and the most likely cause is the wrong data association due to the low constraint on y-direction. The result in WKW dataset shows that P2P ICP, GICP, and full PMM achieve comparable convergence basin. In the Willow dataset, all the algorithms witness a large offset on translation estimation, which is caused by the highly symmetric structure of the small office rooms. Overall, Full PMM demonstrates superior or comparable performance over existing algorithms.

4.6.3.3 Efficiency Analysis

The overall computation time of full PMM is compared against the ICP, P2P ICP, GICP and NDT methods and results are shown in Table 4.4. The results indicate that ICP, P2P ICP, and full PMM require a significantly lower computation time compared to NDT and GICP. The high computation time of NDT is due to computing

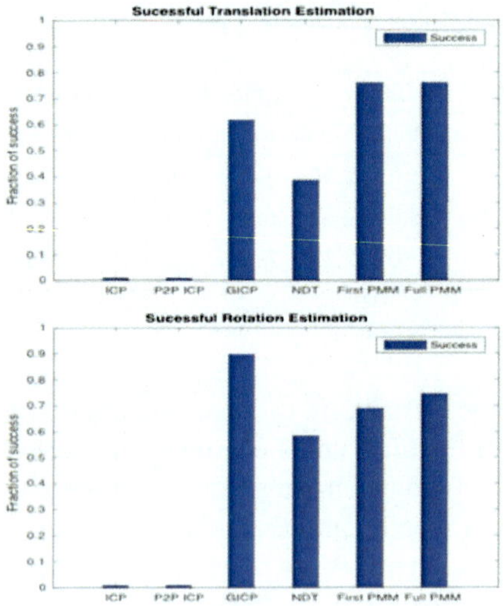

(a) The success rate in Corridor dataset

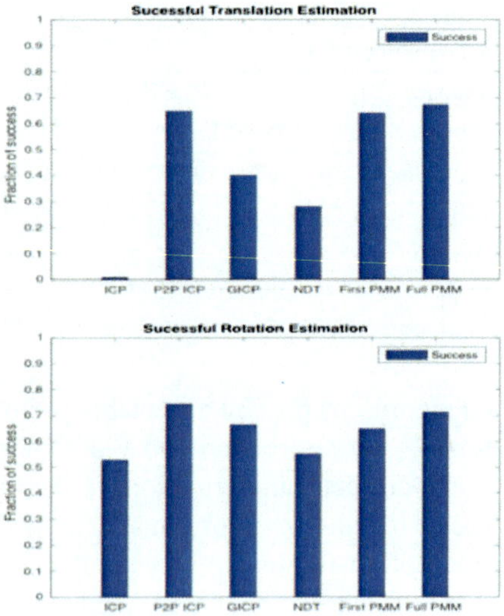

(b) The success rate in WKW dataset

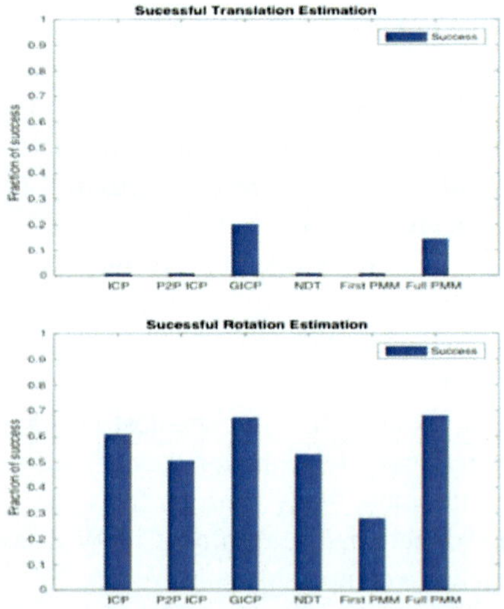

(c) The success rate in Willow dataset

Fig. 4.12 Summary of success rates of the evaluated matching methods over all datasets. Translation estimation (top) and rotation estimation (bottom)

the required normal distributions from the octomap input in addition to the fusion step. GICP adopting a more similar approach to PMM, generates the second best map registration accuracy of all methods, however it's higher computation time can be attributed to the use of only the voxel level registration requiring more computation efforts compared to our proposed two level PMM. Therefore, it can be claimed that the high efficiency of the full PMM compared to all other methods stems from the

Table 4.4 The efficiency comparison of the algorithms (Seconds)

Datasets	ICP	P2P ICP	GICP	NDT	Full PMM
Corridor	3.14	2.90	24.08	136.42	**2.13**
WKW	10.35	8.13	46.66	233.33	**5.31**
Willow	5.59	4.47	40.27	217.38	**3.83**

use of the two-level structure, where edges are registered in the first level with fast convergence with high accuracy which causes the voxel based registration at the second level to converge quickly as well, making the entire process more efficient.

4.6.4 Statistical Testing and Map Merging

In the standard library, GICP and NDT cannot provide point-wise correspondence. We only discuss ICP, P2P ICP and full PMM in the following experiments. After computing the transformation, the accuracy is evaluated by calculating the percentage of inliers and merged based on relative entropy filter.

4.6.4.1 Statistical Testing and Transformation Evaluation

Figure 4.13 illustrates the comparison of the average Mahalanobis distances among P2P ICP map matching, ICP map matching and full PMM in the three experiment settings. The figure shows the box plots that enclose the middle 50% of the data with the median value marked by a line. It is apparent that our algorithm has lower variance and mean compared with others.

Since ICP is deterministic with only coordinate position, the covariance of each point is denoted as the identity matrix $I_{3\times3}$. For the three experiments, a strict criterion is set to guarantee the accuracy: the inliers threshold is set as 0.2 Mahalanobis distance, and the fraction of rejecting a transformation is set according to the overlapping rate. The overlapping rate for Willow dataset, WKW dataset, and Corridor dataset are 40%, 42%, and 48%, then, the acceptance is calculated as 70%, 69%, and 66% respectively.

In Fig. 4.14, the acceptance threshold is denoted with a black line, where the inlier rate higher than the threshold is accepted. In Corridor and Willow datasets, the inlier percentages of ICP and P2P ICP are lower than full PMM since they fail to converge to perfect alignment. Based on the auto-tuned thresholds, full PMM transformation generates accurate and reliable map registration result compared with ICP and P2P ICP based map matching approach. Alternatively, both full PMM and P2P map matching algorithms produce acceptance results in the WKW environment.

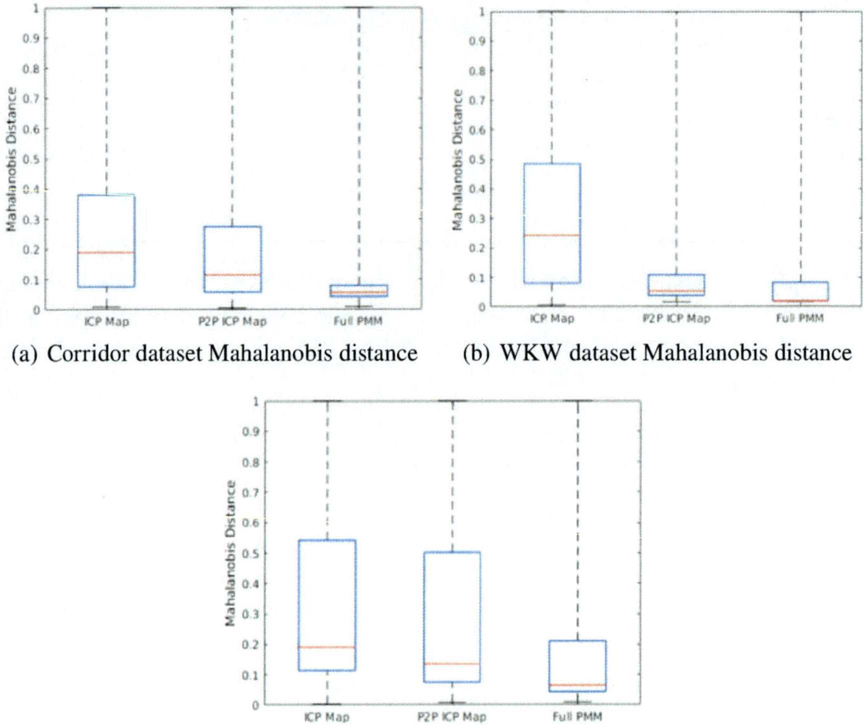

(a) Corridor dataset Mahalanobis distance (b) WKW dataset Mahalanobis distance

(c) Willow dataset Mahalanobis distance

Fig. 4.13 The distribution of the average Mahalanobis distance between the matched pairs in all three experiments

Table 4.5 Average entropy of fused map for different update rule. (The best performance is denoted in bold.)

	Directly Stitch		Taking Average		Relative Entropy	
Dataset	Size	Entropy	Size	Entropy	Size	Entropy
Corridor	40763	0.2061	**32787**	0.2141	**32787**	**0.2002**
WKW	101364	0.2374	**82765**	0.2371	**82765**	**0.2317**
Willow	95596	0.1583	**75442**	0.1680	**75442**	**0.1526**

4.6.4.2 Relative Entropy Filter Based Map Merging

After determining the correct transformation for map registration, a relative entropy filter is applied to achieve map merging by considering the dissimilarities between matched voxels. The relative entropy threshold is set as 0.1 in KL-divergence. The entropy of the resultant maps after applying the KL-divergence based merging algorithm is summarized in Table 4.5. In comparison, the merging strategy in [6] is performed by directly taking the average of probabilities for all the matched voxels.

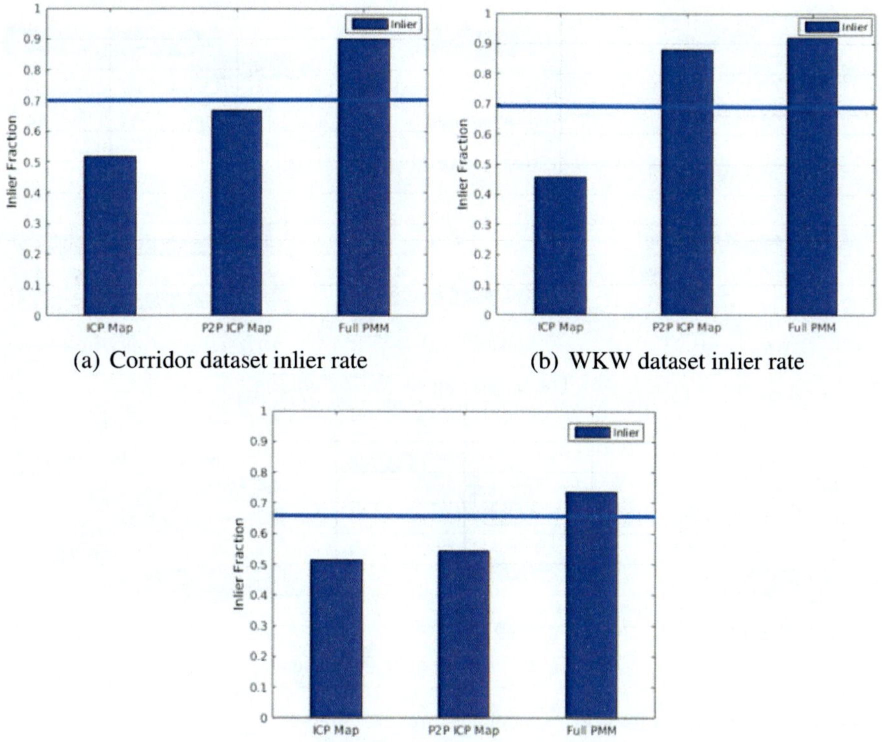

(a) Corridor dataset inlier rate

(b) WKW dataset inlier rate

(c) Willow dataset inlier rate

Fig. 4.14 The fraction of inliers and the threshold of accepting the estimated transformation

Besides, the algorithm is also compared with directly stitching the two maps without any post-processing.

The results are summarized in Table 4.5. It is shown that the relative entropy filter based map merging algorithm reduces the size of the fused map and mean of the entropy. Both relative entropy filter and taking an average algorithm to lead to compression for the size of the map compared with directly stitching. The entropy of WKW and Corridor environment is higher than Willow due to real-world noise and uncertainty. The experiment indicates that the proposed map fusion strategy can combine probabilities of both maps effectively to decrease the total uncertainty. Figure 4.15 shows the final results of the fused 3D occupancy grid maps of three environments.

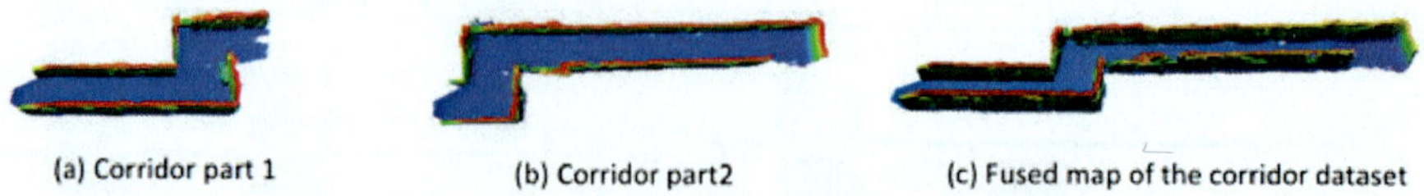

(a) The fused map of Corridor dataset

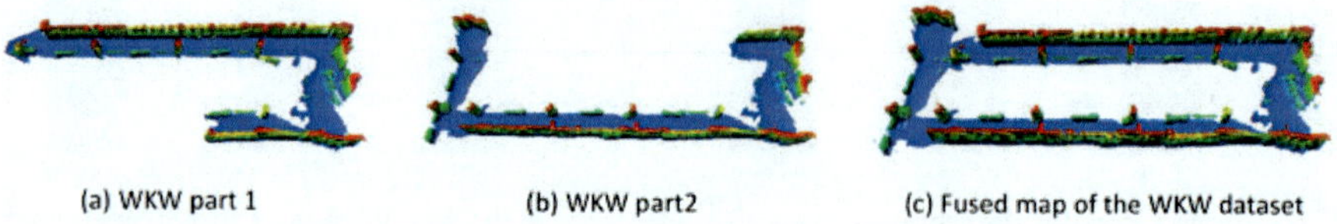

(b) The fused map of WKW dataset

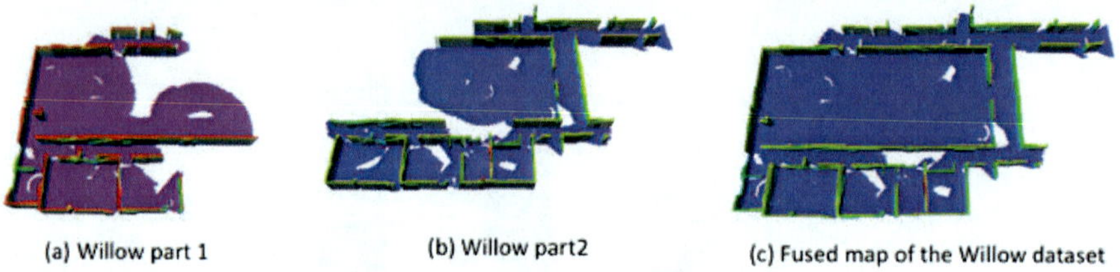

(c) The fused map of Willow dataset

Fig. 4.15 The results of fused maps for all the three experiments

4.7 Conclusions

This chapter extends the works of Chap. 3 by proposing a novel hierarchical probabilistic map fusion framework for 3D occupancy grid maps, where the uncertainties of the structural features and individual voxels are used during matching in a two-step approach. This increases the convergence rate of the optimization process as the use of local uncertainty prevents outliers from dominating the solution.

The work addresses the hierarchical probabilistic fusion framework for 3D occupancy maps, which is factorized into three modules: uncertainty modeling, map matching, and map merging. The chapter first clarifies the difference between raw sensor data registration and map matching in-depth. Then, the local uncertainty of maps is modeled by combing the individual voxel and structural edge covariances. To calculate the accurate relative transformation between maps, a two-level probabilistic map matching algorithm is proposed to combine structural information and voxel information. For the first level, 3D edges extracted from the maps are aligned based on the probabilistic data association strategy. Then, the full maps are correlated on voxel-wise to refine the result and derived the corresponding pairs for probability merging. Since the transformation is only a local minimum, statistical testing is performed to evaluate the quality of the transformation by rejecting low inlier fraction. The final map merging is achieved with a relative entropy filter which integrated the measurements from both maps and decreased the uncertainties. Experiments on simulated and real-world environments are conducted, and the evaluations confirms the efficacy of the proposed approach in generating more accurate relative transformation estimates between maps with a faster convergence rate and a larger convergence basin. The proposed algorithm is able to produce more consistent global 3D occupancy grid maps compared with existing methods.

In this chapter, the map fusion problem in structured environment with a specific sensor type is discussed. For a group of robots in arbitrary (structured/semi-structured/unstructured) environments, single sensor is far from enough in mapping accuracy and robustness. Maps generated with heterogeneous sensors need to be matched and merged to combine information with a higher information content. Chapter 5 will present a general probabilistic framework for collaborative 3D mapping using heterogeneous sensors.

References

1. Besl PJ, McKay ND (1992) Method for registration of 3-D shapes. In: Sensor fusion IV: control paradigms and data structures, vol 1611, pp 586–606. International Society for Optics and Photonics
2. Ying S, Peng J, Du S, Qiao H (2009) A scale stretch method based on ICP for 3d data registration. IEEE Trans Autom Sci Eng 6(3):559–565
3. Segal A, Haehnel D, Thrun S (2009) Generalized-ICP. In: Robotics: science and systems, Seattle, WA, vol 2, p 435
4. Yue Y, Wang D, Senarathne P, Moratuwage D (2016) A hybrid probabilistic and point set registration approach for fusion of 3d occupancy grid maps. In: 2016 IEEE international conference on systems, man, and cybernetics (SMC), IEEE, pp 001975–001980
5. Rusu RB, Cousins S (2011) 3d is here: Point cloud library (PCL). In: 2011 IEEE international conference on robotics and automation, IEEE, pp 1–4
6. Jessup J, Givigi SN, Beaulieu A (2015) Robust and efficient multirobot 3-d mapping merging with octree-based occupancy grids. IEEE Syst J 11(3):1723–1732

Chapter 5
Collaborative 3D Mapping Using Heterogeneous Sensors

Operating multiple robots in an unstructured environment is challenging due to its high complexity and uncertainty. In such applications, the integration of individual maps generated by heterogeneous sensors is a critical problem, especially the fusion of sparse and dense maps. This chapter proposes a general multi-level probabilistic framework to address the integrated map fusion problem, which is independent of sensor type and SLAM algorithm employed. The key novelty of this chapter is the mathematical formulation of the overall map fusion problem and the derivation of its probabilistic decomposition. The framework provides a theoretical basis for computing the relative transformations amongst robots and merging probabilistic map information. Since the maps generated by heterogeneous sensors have different physical properties, an expectation-maximization-based map-matching algorithm is proposed which automatically determines the number of voxels to be associated. Then, a time-sequential map merging strategy is developed to generate a globally consistent map. The proposed approach is evaluated in various environments with heterogeneous sensors, which demonstrates its accuracy and versatility in 3-D multirobot map fusion missions.

5.1 Introduction

In Chap. 4, high-level structural and low-level voxel features are combined for map matching. However, they only focus on homogeneous sensors in structured environment. The next challenge is to extend these techniques to multi-robot systems equipped with miscellaneous sensors in unstructured environment.

In single robot level, with the emergence of diverse sensor types, each robot could carry heterogeneous sensors(laser scanners and vision-based sensor). The nature of

Y. Yue and D. Wang, *Collaborative Perception, Localization and Mapping for Autonomous Systems*, Springer Tracts in Autonomous Systems 2,
https://doi.org/10.1007/978-981-15-8860-0_5

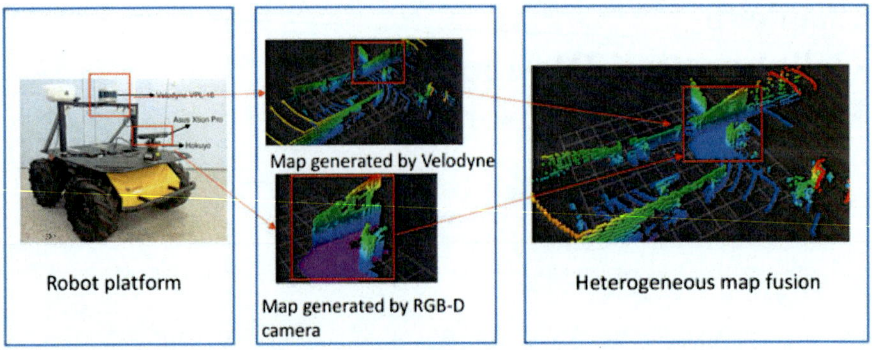

Fig. 5.1 An example of the heterogeneous map fusion

the map produced by different sensors is the major challenge, i.e, density, range, sensor noise and field of view. Considering the 3D Lidar (Velodyne) scans 360 degree and records most of the environment without leaving any major blind spots, however, generates a less dense map owing to sensor sparseness. While the RGB-D camera (Asus Xtion) generates denser and more detailed 3D map with the surroundings, however, RGB-D camera only has limited field of view and the error grows exponentially with the increasing of the sensing range. Hence, the fusion of maps produced by miscellaneous sensors could present more detailed information of the environment and increase the perception ability of the robot (See Fig. 5.1. Left: Robot platform Husky A200 equipped with various sensors. Middle: The map generated by Velodyne and RGB-D camera. Right: fused global map.).

In multi-robot level, considering the complexity and dimension of the environment, each robot only has partial filed of view and limited sensing ability of the surroundings. In order to allow the team of robots to model the environment and plan tasks from a systematic viewpoint, sharing and fusing map data perceived by each robot among all robots act as the basis for multi-robot coordination. The research of exploiting multi-robot map fusion framework in both single robot and multi-robot levels has not been studied extensively [1]. In general, map fusion on both levels is composed of estimating the relative transformation and integrating the partial map information, which are usually tackled separately [2]. Existing approaches have relied on low level point set registration based algorithms for map matching [3], where hard one-to-one correspondences relationship are assigned. When maps produced by heterogeneous maps need to be matched, it is hard to detect correct corresponding points [4]. Most researchers focus on estimating relative transformation and directly applying the transformation to stitch the partial maps [5], where the integrating of transformation uncertainty and map uncertainty are not considered. In addition, they have not provided a comprehensive analysis of the integrated map fusion system on map uncertainty modeling, probabilistic map matching, transformation evaluation and accurate map merging.

In this chapter, a novel multi-level framework is presented allows for joint local and global map matching and merging, which acts as the theory basis for multi-robot mapping missions. More specifically, the multiple data association method to match heterogeneous maps; Then a time-sequential map merging algorithm is applied. We designed an efficient distributed solution and perform all computation online and on board to gain robustness w.r.t. periodic communication and robot failure. The main contributions of this chapter are listed as follows:

- A general probabilistic framework is proposed to address the integrated map fusion problem, which is independent of sensor types and SLAM algorithms. The framework provides theoretical basis for computing the relative transformations (map matching) and merging probabilistic map information (map merging).
- A probabilistic map matching algorithm is proposed that automatically decides the numbers of voxels to be associated for maps produced by heterogeneous sensors with different physical properties.
- A time-sequential map merging framework that makes fusing maps from distributed multi-robot system efficiently, where uncertainty embedded in the map and caused by the relative transformation are propagated and integrated.

5.2 Distributed Multi-Robot Map Fusion

In this section, the system architecture for multi-robot map fusion with single robot level and multi-robot level is presented, where the problem is formulated to provide a theoretical basis for its factorization into two sub problems of map matching and map merging.

5.2.1 System Architecture

The overview of the system architecture is depicted in Fig. 5.2. The focus of this chapter is to build a general framework for map fusion, which can be divided into two sub-problems, i.e., map matching and map merging. **First level** is perception level, the robot equipped with heterogeneous sensors will generate its map. For example, laser scanner tends to generate 360 degree sparse map, while rgb-d camera generates denser map with narrow field of view. **Second level** is local map level, the maps produced by different sensors are fused locally on each robot to combine the strength for each sensors. **Third level** is global map level, the robots communicate with each other to transmit and fuse the local maps into a global map. Our main focus is on the second level and third level.

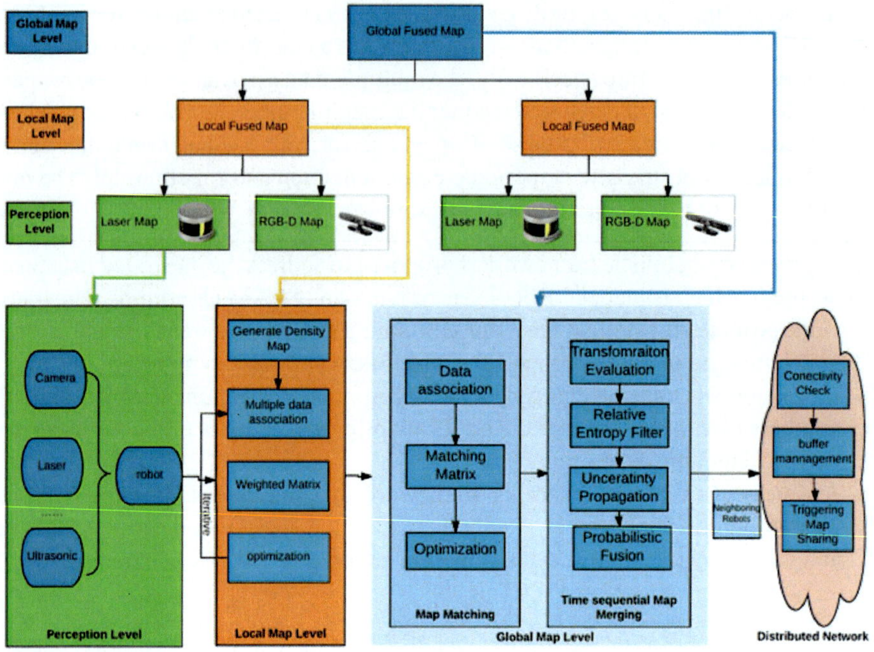

Fig. 5.2 The diagram of the system

5.2.2 *System Framework Definition and Formulation*

Problem 1 *Given a group of robots $\mathcal{R} = \{r\}_{1:n_r}$, each robot r is equipped with a set of heterogeneous sensors $\mathcal{S} = \{s\}_{1:n_s}$, the map produced by robot r using sensor s is denote as $m_{r,s}$, the problem is to estimate the fused global map M_r and the set of relative transformation $\mathcal{T}_r$ with a fully distributed network.*

Let $\mathcal{R} = \{r_1, r_2, \ldots, r, \cdots\}$ (the number of robots is n_r) be the set of robots in an environment. The neighboring robots of robot r is defined as $\mathcal{R}_r = \{r_n, r_n \in \mathcal{R}_r\}$, where neighboring robots $\mathcal{R}_r$ is the set of robots in communication with r and $\mathcal{R} = \{r \cup \mathcal{R}_r\}$ if the communication covers all robots. A diagram of group of robots is shown in Fig. 5.3.

In the system, input is the perception level and output is the global map level. Local map level is an intermediate level that acts as communication nodes and single robot fusion master. The benefits of introducing the local map level comes at following advantages, firstly: the size of the fused local map is smaller than the summation of individual sensor maps, which reduce the communication burden; secondly, the fused map on single robot level helps the localization and exploration of single robot, where the maps produced by different sensors are fused to compensate the limitation of each types of sensor; thirdly, map fusion happens at perception level brings the computational complexity of $O(R^2S^2)$, while the local map level reduces it to $O(R^2 + RS^2)$. Then, the multi-level framework is factorized into estimating

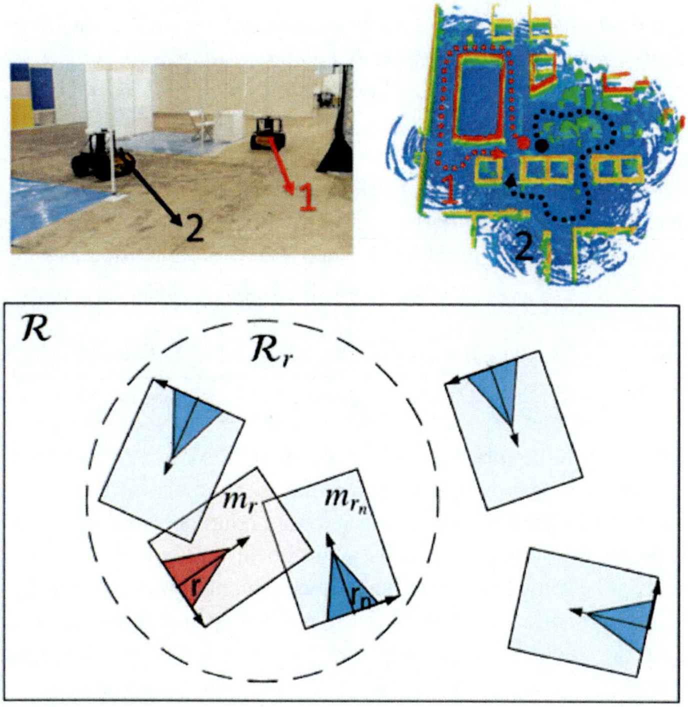

Fig. 5.3 Top: Robot platform and the fused 3D map created using the two robots. Bottom: The diagram of the group of robots

local fused map M_r given the maps of perception level $m_{r,s}$ and estimating fused global map $M_{\mathcal{R}_r}$ conditional on available local maps M_r.

$$\underbrace{\prod_{r\in\mathcal{R}} p(m_r, T_{S_r}|m_{r,1:n_s})}_{\text{local map level}} \cdot \underbrace{p(M_r, T_{r,\mathcal{R}_r}|m_r, m_{\mathcal{R}_r})}_{\text{global map level}} \tag{5.1}$$

For local map level, each individual robot r is assumed to estimate its trajectories based on local observations and generates its partial map m_r^s, then robot r fuses the sensor maps $m_r^{1:n_s}$ to generate M_r. For global map level, robot r receives the set of local maps $M_{\mathcal{R}_r}$ from nearby robots, where $M_{\mathcal{R}_r} = \{M_{r_n}, r_n \in \mathcal{R}_r\}$.

5.2.3 Map Fusion Definition and Formulation

Problem 2 *For both local and global robot levels, the problem of map fusion problem is an joint probability estimation of relative transformation and merged map.*

Assuming the relative transformation estimation and map merging are conditionally independent, the map fusion problem is factorized into estimating a set of relative transformation (map matching) and generating the global map conditioned on all available partial maps (map merging).

To approximate the joint distribution, both local and global map fusion in Eq. (5.1) can be factorized into a product of relative transformation posterior and the fused map posterior. To simply the derivation, only global map level is derived in detail, where it is the same for the local level. Then, Eq. (5.1) is written as follows:

$$p(M_r^{\mathcal{R}_r}, T_{r,\mathcal{R}_r}|m_r, m_{\mathcal{R}_r}) = \underbrace{p(T_{r,\mathcal{R}_r}|m_r, m_{\mathcal{R}_r})}_{\text{map matching}} \cdot \underbrace{p(M_r^{\mathcal{R}_r}|T_{r,\mathcal{R}_r}, m_r, m_{\mathcal{R}_r})}_{\text{map merging}} \tag{5.2}$$

This decouples multi-robot map fusion problem into two separate estimation problems. The key advantage of this factorization is summarized as follows: Firstly, it benefits from using MAP estimation for robots relative transformation estimation, making it possible to apply probabilistic nonlinear least square minimization models. Secondly, it benefits from utilizing a time-sequential relative entropy filter based map merging algorithm. Thirdly, the output of map matching result can be used to propagate the uncertainty for map merging, which tightly concatenates the two sub-problems.

5.3 Multi-Robot Map Matching

In this section, a general framework is presented for map matching. To address the challenge of matching maps produced by heterogeneous sensors, the estimation of relative transformation based on expectation maximization is proposed.

5.3.1 Mathematic Formulation

For robot r, it receives partial maps sequentially, which means the relative transformation is a serial processing upon receiving of new partial maps. Since the relative transformation T_{r,r_n} is also pairwise independent of each pairs of robots, the posterior of relative transformation $p(T_{r,\mathcal{R}_r}|m_r, m_{\mathcal{R}_r})$ can be factorized into a product of the posterior of relative transformation between each pair, as shown in Eq. (5.3):

$$p(T_{r,\mathcal{R}_r}|m_r, m_{\mathcal{R}_r}) = \prod_{r_n \in \mathcal{R}_r} p(T_{r,r_n}|m_r, m_{r_n}) \tag{5.3}$$

where $T_{r,r_n} \in SE(3)$ is a rigid transformation with rotation $R_{r,r_n} \in SO(3)$ and translation $t_{r,r_n} \in R^3$. Instead of simply considering T_{r,r_n} as a fixed unknown parameter, it is modeled as a 6D random variable. Then, T_{r,r_n} can be modeled as a multivariate Gaussian distribution and solved by MAP estimation, where $T_{r,r_n} \sim (\hat{T}_{MAP}, \hat{T}_{\Sigma_{MAP}})$. To solve the MAP estimation problem, Bayes rule is applied to factorize Eq. (5.3) into Eq. (5.4), which is the product of maximum likelihood estimation and prior estimation:

$$p(T_{r,r_n}|m_r, m_{r_n}) = p(m_r|T_{r,r_n}, m_{r_n}) \cdot p(T_{r,r_n}), \tag{5.4}$$

where term $p(m_r|T_{r,r_n}, m_{r_n})$ is a common maximum likelihood estimation (MLE) problem that aims to find the most likely relative transformation T_{r,r_n} by matching the two partial maps m_r and m_{r_n}. By specifying the initialization, prior information is incorporated to evaluate the posterior even with a poor initial input, which can be given by feature descriptor correlation. If no prior information is available, the prior distribution is assumed to be a uniform distribution. In this case, $\hat{T}_{MAP} = \hat{T}_{MLE}$, where MLE is a special case of MAP estimation. To establish voxel-wise correspondences for map merging, the dense scan-matching technique is applied to solve MLE.

5.3.2 3D Occupancy Map Matching

For dense registration of two maps m_r and m_{r_n}, the common solution is to fix the map of current robot m_r as model map, and the neighboring map m_{r_n} as scene map, where the rigid transformation T_{r,r_n} is estimated by transforming m_{r_n} to the coordinate of m_r and minimizing the error metric iteratively. To estimate T_{r,r_n}, data association needs to be estimated firstly, which is usually addressed by performing nearest neighborhood search of Euclidean distance or Mahalanobis distance. For pure registration purpose, the matches are not parameters of interest and are usually ignored after registration.

In our previous work [6], a novel map registration algorithm is applied called occupancy iterative closest point(OICP), which establishes voxel-to-voxel correspondences between all overlapped partial maps. Here, the data association issue is further exploited with different cases. For the partial map, $m_r = \{m_r^i\}_{i=1}^{N_{m_r}}$, $m_{r_n} = \{m_{r_n}^j\}_{j=1}^{N_{m_{r_n}}}$, where N_{m_r} and $N_{m_{r_n}}$ are the number of voxels of partial maps. Since the voxel-wise correspondences are usually assumed to be independent, the probability density function of $p(m_r|m_{r_n}, T_{r,r_n})$ can be factorized into Eq. (5.5).

$$p(m_r|m_{r_n}, T_{r,r_n}) = \prod_i p(m_r^i|m_{r_n}^{c_i}, T_{r,r_n}) \tag{5.5}$$

where $m_{r_n}^{c_i} = \{m_{r_n}^j, j \in c_i\}$ is the set of candidate corresponding voxels of m_r^i. The problem of calculating the corresponding points $m_{r_n}^{c_i}$ of m_r^i is often referred as data association. To estimate the hidden data association, Expectation-maximization (EM) is introduced, which is an algorithm well-suited for models containing latent variables and estimating them iteratively by marginalizing the likelihood function.

There are two steps for EM: the E-step efficiently estimates the hidden variables by evaluating the expectation, while the M-step solves the rigid transformation based on E-step using optimization algorithms. Here, we define a binary matching matrix J to represent the estimation of hidden data association between the corresponding voxels, where $J_{i,j} = 1$ if m_r^i corresponds to $m_{r_n}^j$ and equals 0 otherwise. For each row i, the matched voxels of m_r^i is the set of $J_{i,c_i} = \{J_{i,j} = 1, j \in c_i\}$. The EM form of Eq. (5.5) is rewritten as:

$$
\begin{aligned}
p(m_r|m_{r_n}, T_{r,r_n}) &= \sum_J p(m_r, J|m_{r_n}, T_{r,r_n}) \\
&= \prod_i \sum_{c_i} p(m_r^i, J_{i,c_i}|m_{r_n}^{c_i}, T_{r,r_n}) \\
&= \prod_i \sum_{c_i} \underbrace{p(J_{i,c_i}|m_r^i, m_{r_n}^{c_i}, T_{r,r_n})}_{\text{E step:data association}} \cdot \underbrace{p(m_r^i|J_{i,c_i}, T_{r,r_n}, m_{r_n}^{c_i})}_{\text{M step:optimization}}
\end{aligned} \tag{5.6}
$$

The negative log-likelihood to be minimized with EM is given by

$$
f(T_{r,r_n}) = \min_{T_{r,r_n}} \sum_i \sum_{c_i} -p(J_{i,c_i}|m_r^i, m_{r_n}^{c_i}, T_{r,r_n}) \cdot \ln p(m_r^i|J_{i,c_i}, T_{r,r_n}, m_{r_n}^{c_i}) \tag{5.7}
$$

5.3.3 E-Step

- **Homogeneous Sensors**: Since the density of map is similar for the same type of sensors, one-to-one hard assignment is established for computational efficiency. For each voxel m_r^i, there will be one and only one voxel $m_{r_n}^j$ correspond to it, where $J_{i,c_i} = J_{i,j}$.

$$
\forall m_r^i \in m_r, \exists! m_{r_n}^j \in m_{r_n} \Rightarrow J_{i,j} = \omega_{ij}
$$

- **Heterogeneous Sensors**: If heterogeneous sensors need to be aligned, one-to-one matching might fail due to wrong matches or insufficient correspondences. Hence, multiple soft correspondences should be established, where each voxel m_r^i corresponds to multiple voxels in $m_{r_n}^{c_i}$

$$
\forall m_r^i \in m_r, \exists m_{n_c r}^{c_i} \in m_{r_n} \Rightarrow J_{i,c_i} = \{J_{i,j} = \omega_{ij}, j \in c_i\}
$$

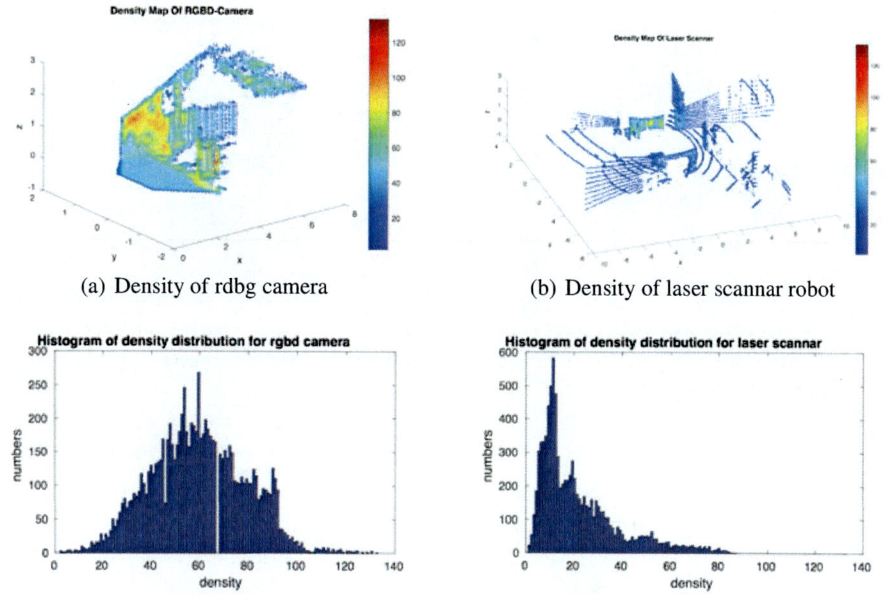

(a) Density of rdbg camera

(b) Density of laser scannar robot

(c) Density historgram of rdbg camera

(d) Density historgram of laser

Fig. 5.4 The density map and histogram for the two sensors

The problems comes at two hands: (1) automatically determining the policy for the candidate correspondences searching. If it is the case of heterogeneous sensors, how to determine the value of c_i to balance the efficiency and accuracy. (2) For heterogeneous sensors, the candidate $J_{i,j} \in J_{i,c_i}$ should have different weight to contribute to the objective function.

As can been in Fig. 5.4, the density of rgb-d camera is higher than laser scanner. In addition, the histogram of rgb-d map complies a Gaussian distribution, which means the density is distributed more evenly. However, the histogram of laser is more like the Gamma distribution. Based on the observation, a local density map based automatic tuning strategy is proposed for probabilistic data association, as shown in Algorithm 3. For the given maps, the algorithm firstly generates their density maps (depicted in Fig. 5.4. If the mean of the density maps are similar, one-to-one correspondence is applied. If not, the numbers of correspondences (c_i) to be established are determined by the ratio of local density value. As shown in Fig. 5.5, the density of rgbd-d map is distributed evenly, the numbers of voxels matched increases with the decreases of laser map density. Each voxel in laser map matches different value to c_i in rgb-d map, where the histogram and density map shows the automatically tuned value of c_i.

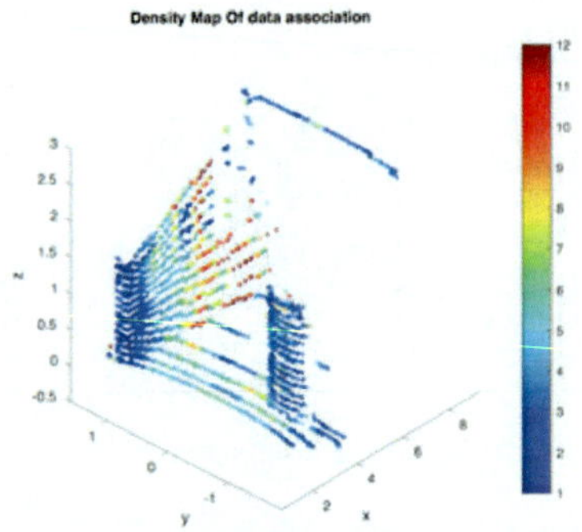

(a) Numbers of voxels matched in rgb-d map for each voxel in laser map

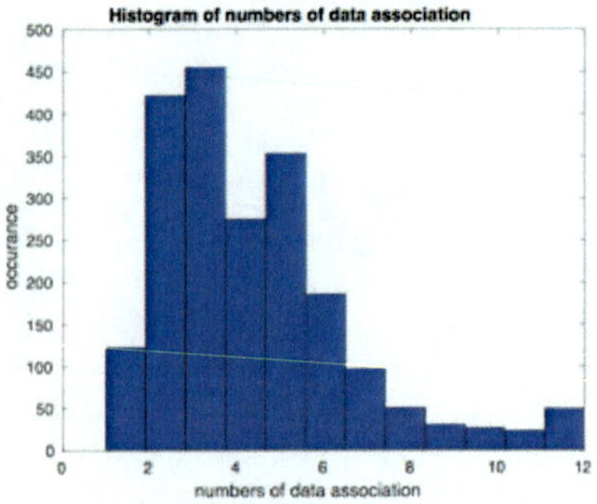

(b) Histogram of numbers of voxels matched

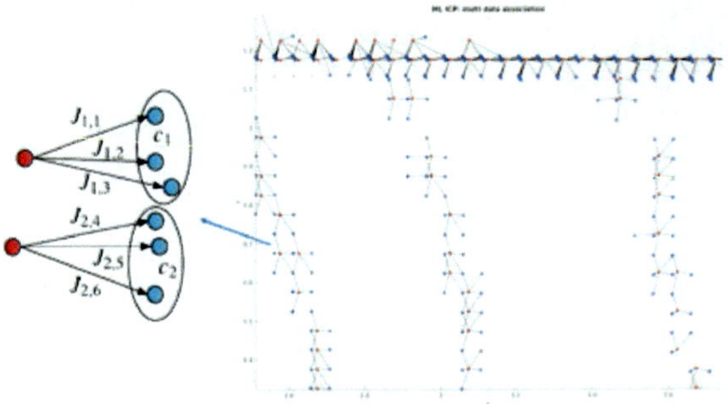

(c) The edges show the multiple data association established. Red points represent laser map, while blue points are rgb-d map.

Fig. 5.5 Density based multiple data association strategy

5.3.4 M-Step

Solving M-step is to minimize the log-likelihood of Eq. (5.6) based on the data association established in E-step. Here, probabilistic weights are assigned to each pair of correspondences. There are several different ways of updating the rotation and translation, for example, Levenberg-Marquardt, single value decomposition.

$$f(T_{r,r_n}) = \min_{T_{r,rn}} \sum_{i} \sum_{j}^{c_i} -\omega_{i,j} \cdot \ln p(m_r^i | J_{i,j}, T_{r,r_n}, m_{r_n}^j) \qquad (5.8)$$

5.4 Time-Sequential Map Merging

After the relative transformation is derived, the partial map m_{r_n} is transformed to coordinate frame of m_r by transformation function $\Phi(T_{r,r_n}, m_{r_n})$, to generate ${}^r m_{r_n} = \Phi(T_{r,r_n}, m_{r_n})$. The process of combing the information of common objects from

Algorithm 3 Probabilistic Map Matching

Require: The maps $m_r = \{m_r^i\}_{i=1}^{N_{m_r}}$, $m_{r_n} = \{m_{r_n}^j\}_{j=1}^{N_{m_{r_n}}}$, and initial transformation fuction T_0
Ensure: Iteratively minimize the error function
  1. generate density map d_r for m_r and d_{r_n} for m_{r_n}
  for k from 1 to termination criteria **do**
    for i from 1 to N_{m_r} criteria **do**
      if $round(\frac{mean(d_r)}{mean(d_{r_n})}) = 1$ **then**
        2. one-to-one matching, where $J_{i,c_i} = J_{i,j}$
      else
        3. multiple data association, where $c_i = round(\frac{d_r^i}{d_{r_n}^j})$
      end if
      4. assign weight: $\omega_{ij} = \exp^{-(T^k * m_r^i - m_{r_n}^j)}$, $j \in c_i$
      5. iteratively optimize the error function: $\hat{T}_{r,r_n}^{k+1} = argmin \sum_i \sum_j^{c_i} \omega_{i,j} \cdot \ln p(m_r^i, J_{i,j} | T_{r,r_n}^k, m_{r_n}^j)$
    end for
  end for

partial maps to form an enhanced global map is usually referred to as map merging and is formulated in Eq. (5.9):

$$p(M_r^{\mathcal{R}_r} | T_{r,\mathcal{R}_r}, m_r, m_{\mathcal{R}_r}) = p\left(M_r^{\mathcal{R}_r} | m_r, \Phi_{r_n \in \mathcal{R}_r}(T_{r,r_n}, m_{r_n})\right) \tag{5.9}$$

During real missions, maps are generated and transmitted sequentially. This results in robot r receiving its neighboring robots maps in some permutation π of $\mathcal{R}_r$. Hence, the map merging can be performed serially, where reception of a new partial map would trigger a pair-wise map merging. Then, Eq. (5.9) can be factorized into Eq. (5.10).

$$p(M_r^{\mathcal{R}_r} | m_r, \Phi_{r_n \in \mathcal{R}_r}(T_{r,r_n}, m_{r_n})) = \prod_{r_n=1}^{N_r} p(M_r^{\pi(1,r_n)} | M_r^{\pi(1,r_n-1)}, \Phi(T_{r,\pi(r_n)}, m_{\pi(r_n)})) \tag{5.10}$$

Note that $M_r^{\pi(1,r_n)}$ is the partial global map of robot r generated by fusing the maps of robots $\pi(1)$ to $\pi(r_n) \in \mathcal{R}_r$ with m_r after the reception of map $m_{\pi(r_n)}$ from neighboring robot $\pi(r_n) \in M_r^{\mathcal{R}_r}$. The final global fused map $M_r^{\mathcal{R}_r}$ of robot r, can be retrieved after the end of this serial process, hence $M_r^{\mathcal{R}_r} = M_r^{\pi(1,Nr)}$.

5.4.1 Uncertainty Propagation and Transformation

In general, there are two types of uncertainty need to be integrated when performing map merging.

(1) *Map uncertainty*: in real environment, uncertainties will occur due to localization error and sensor measurement noise. In map level, those uncertainties are represented with the occupancy probability value. Since the same object can be observed in different perspectives by different robots, voxels representing this object could have different occupancy probabilities in separated maps. Hence, it is vital to consider these occupancy probability dissimilarities when fusing them into a global map.

(2) *Relative transformation uncertainty*: the error of the robots relative position estimation. This uncertainty is related to the estimation results, the more accurate the estimation the smaller the uncertainty. This type of uncertainty also needs to be considered in map merging process. The relative estimation uncertainty is propagated by the map matching algorithm and will be represented as a covariance matrix.

Our approach is to propagate the uncertainty for both single robot and multi-robot level, then the corresponding probability distributions are merged to output an improved estimate of the observed object. The assumption for the mathematic is based on that sensor errors are independent and complies a normal distribution. Assume that $q = [q_x \ q_y \ q_z \ L_q]$ denotes the position of the extracted voxel center $[q_x \ q_y \ q_z]$ and the occupancy probability L_q. The mean of q is $\mu_q = [\mu_{q_x} \ \mu_{q_y} \ \mu_{q_z}]$, the covariance is a diagonal matrix with the same entry $\Sigma_q = diag[\sigma^2_{q_x q_x} \ \sigma^2_{q_y q_y} \ \sigma^2_{q_z q_z}]$. Voxel q is transformed by the following function to voxel r.

$$r = f(\alpha, \beta, \gamma, t_x, t_y, t_z, q_x, q_y, q_z) = Rq + t \tag{5.11}$$

where α, β, γ are the rotation angles and t_x, t_y, t_z are the translation elements. Since the transformation function $f(\cdot)$ is nonlinear, then the transformed voxel r will be non-gaussian. To approximate r with Gaussian, the transformation $f(\cdot)$ is linearized using the first-order Taylaor expansion:

$$p(r) = \frac{1}{\sqrt{det(2\pi\Sigma)}} \exp\left\{-\frac{1}{2}(r - \mu_r)^T \Sigma_r^{-1} (r - \mu_r)\right\} \tag{5.12}$$

$$\sim \mathcal{N}(r; \mu_r, \Sigma_r) \tag{5.13}$$

where μ_r can be directly calculated using Eq. (5.11). Assume the covariance of transformation is $\Sigma_f = diag[\sigma^2_{t_x t_x} \ \sigma^2_{t y t_y} \ \sigma^2_{t_z t_z} \ \sigma^2_{\alpha\alpha} \ \sigma^2_{\beta\beta} \ \sigma^2_{\gamma\gamma}]$. Through deriving the Jacobian of transformation function, the covariance of transformed voxel Σ_r is estimated in Eq. (5.14).

$$F_{\alpha\beta\gamma \ t_x t_y t_z} = \frac{\partial f(\cdot)}{\partial(\alpha, \beta, \gamma, t_x, t_y, t_z)}, \ F_{q_x q_y q_z} = \frac{\partial f(\cdot)}{\partial(q_x, q_y, q_z)}$$

$$\Sigma_r = \underbrace{F_{\alpha\beta\gamma \ t_x t_y t_z} \Sigma_f F^T_{\alpha\beta\gamma \ t_x t_y t_z}}_{\text{transformaiton uncertainty}} + \underbrace{F_q \Sigma_q F_q^T}_{\text{map uncertainty}} \tag{5.14}$$

5.4.2 Uncertainty Merge

We employ our previous work on map merging [6] to execute the sequential pair-wise uncertainty merging. For the two types of uncertainty, the merging process should preserve all valuable information of the partial maps while decreasing the uncertainty of the fused map. Once unidirectional alignment of maps is performed in the map matching stage to establish the correspondence, a relative entropy filter based on Kullback Leibler divergence is applied to merge map measurements that decreases the total uncertainty of the global map.

We apply a three-dimensional statistical approach, where any given multi-dimensional Gaussian distributions can be integrated by applying matrix operations. Let $\mathcal{N}_x(\mu, \Sigma)$ denote the density function of voxel x, then,

$$\mathcal{N}_c(\mu_c, \Sigma_c) \sim \mathcal{N}_r(\mu_r, \Sigma_r) \cdot \mathcal{N}_p(\mu_p, \Sigma_p) \tag{5.15}$$

where,

$$\mu_c = ({\Sigma_a}^{-1} + {\Sigma_b}^{-1})^{-1}({\Sigma_r}^{-1}\mu_r + {\Sigma_p}^{-1}\mu_p) \tag{5.16}$$

$$\Sigma_c = ({\Sigma_r}^{-1} + {\Sigma_p}^{-1})^{-1} \tag{5.17}$$

Then the steps are extended to the entire voxels in the map, where Algorithm 4 and Fig. 5.6 show the overall process. Firstly, the source voxel q and target voxel p is represented as a 3D gaussian distribution. In line 2–4, q is transformed to r according to the given transformation function T, where covariance of transformed voxel is calculated using Eq. (5.14). Then the final merging happens in line 5 and the uncertainty of both source and target map is fused using Eq. (5.15). The results shows the merging process decreases the uncertainty, and the merged voxel is closer to the one with less uncertainty.

Algorithm 4 Uncertainty Propagation and Map Merging

Require: The maps m_r and $m_{r'}$, and transformation fuction T
Ensure: Transform the map m_r and fuse them
for all voxels q **do**
  1. represent the source voxel as multi-gaussian distribution: $\mathcal{N}_q(\mu_q, \Sigma_q)$
  2. transform the voxel q to r: $\mu_r = T(\mu_q)$
  3. calculate Σ_r using Eq. (5.14)
  4. represent the transformed voxel r with $\mathcal{N}_r(\mu_r, \Sigma_r)$
  5. merge r with corresponding voxel p to get voxel c: $\mathcal{N}_c(\mu_c, \Sigma_c) \sim \mathcal{N}_r(\mu_r, \Sigma_r) \cdot \mathcal{N}_p(\mu_p, \Sigma_p)$

end for

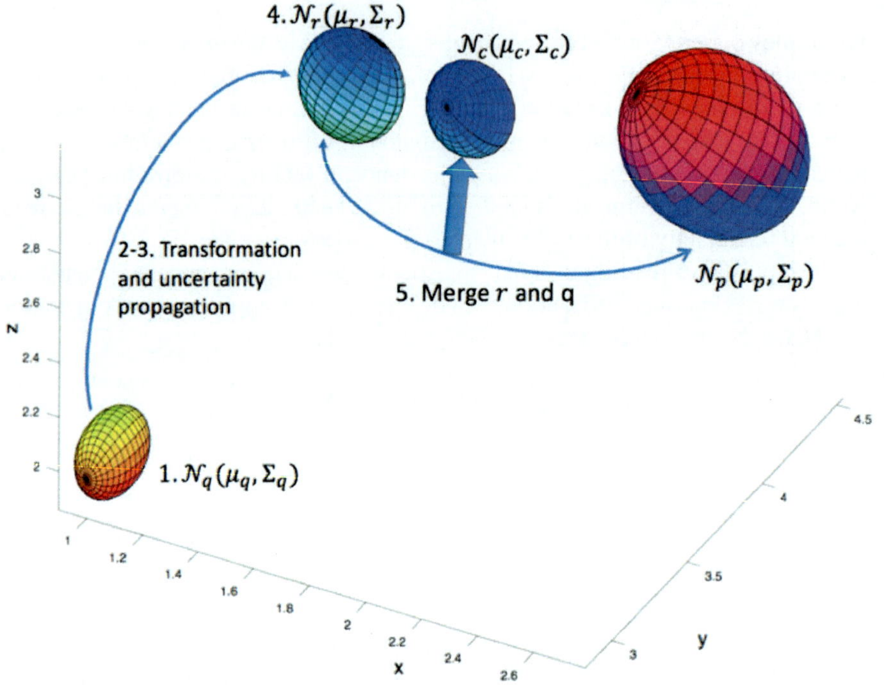

Fig. 5.6 The process of uncertainty propagation and merging

5.5 Experimental Results

The performance of the proposed multi-robot map fusion framework is validated through extensive real world experiments. Both qualitative and quantitative results are presented. The former are obtained by visual inspection of the fused map in different experiment settings. Then, the latter are expressed by calculating the root mean square error and the running time.

5.5.1 Evaluation Protocol

Experiment Environment: In this chapter, four real environment experiments are conducted, where two robots are equipped with heterogeneous sensors. The resolution of the 3D occupancy grid map was set to be 0.1m. The communication between robots was established by long range Wi-Fi with limited bandwidth The experiment settings are:

(a) Indoor Environment (b) Mixed Environment

(c) Changi Centre (d) Unstructured Environment

Fig. 5.7 The experiment setting up: Indoor Environment, Mixed Environment, Changi Centre, Unstructured Environment respectively

(a) Indoor Environment : two Husky robots equipped with ASUS Xtion Camera and & Velodyne 3D Lidar in an indoor environment of Nanyang Technological University.

(b) Mixed Environment : two Husky robots equipped with ASUS Xtion Camera and & Velodyne 3D Lidar in the indoor & outdoor mixed environment of Nanyang Technological University.

(c) Changi Centre: two Husky robots equipped with Velodyne 3D Lidar during Asia Unmanned Exhibition in Changi Centre.

(d) Unstructured Environment: two Husky robots equipped with Velodyne 3D Lidar in an open forest of Nanyang Technological University.

Figure 5.7 and Table 5.1 summarize the setting and the main characteristics of each experiment.

Error Metric: The error metric applied is MSE (Mean Squared Error), which is the mean squared Euclidean distance between the matched pairs.

Comparison Baseline: For each experiment, two matching algorithms are tested. First, the rigid transformation is computed with fixed one-to-to data association correspondence approach, i.e, Occupancy Iterative Closet Point proposed in

Table 5.1 Summary of the experiment environment

Dataset	Sensor	Location	Dynamics	Type
Indoor Environment	Asus Xtion&Velodyne	Indoor	Dynamic	Structured
Mixed Environment	Asus Xtion&Velodyne	Indoor&Outdoor	Static	Semi-Structured
Changi Centre	Velodyne	Indoor	Dynamic	Semi-Structured
Unstructured Environment	Velodyne	Outdoor	Static	Unstructured

Chap. 3. Second, the proposed probabilistic map fusion algorithm is applied to compare the results.

5.5.2 Indoor Environment

The indoor environment is the second floor of a building in the university which contains challenging objects such as reflective glasses and moving people. In order to test the algorithm flexibility of using different sensors with varying characteristics, two robots were tele-operated to explore the environment with different sensors (Asus Xtion PRO and Velodyne VLP-16). The map fusion is implemented at two levels. First, the dense map produced by Asus Xtion and sparse map produced by Velodyne are fused to generate the local fused map. Then, the local fused map was sent to the neighbor robot to generate the global fused map.

In the local level, the Velodyne map is very sparse in the initial position (See Fig. 5.8a), while the Asus Xtion produced dense map at the same time. The matching results of the sparse map and dense map presented in Fig. 5.8a illustrates the superiority of the multiple data association strategy. At $t = 0$, each voxel in the sparse map correlates average 3.47 voxels in the dense map to establish more robust correspondences. With the increasing density of the Velodyne map, the average correlated voxels dropped to 2.15 at $t = 20s$ and two algorithms produce comparable results. In summary, despite the heterogeneity of sensors and challenging environments, the proposed method is able to produce a more consistent map which merges the sparse 3D LiDAR map with the dense RGBD map (see Fig. 5.8b).

In the global map, the local map produced by two robots are fused into a global consistent map. As highlighted in Fig. 5.9 (top part), the RGB-D sensor did not fully perceive the environment at some corner of the building (circled area). However, it still generated a much denser map compared to the sparse 3D LiDAR. Considering the LiDAR scans 360°, it records most of the environment without leaving any major blind spots, however, generates a less dense map owing to sensor sparseness. As shown in Fig. 5.9 (Top: individual map produced by heterogeneous sensors. Middle: the locally fused map by two different robots. Bottom: the global map by fusing two

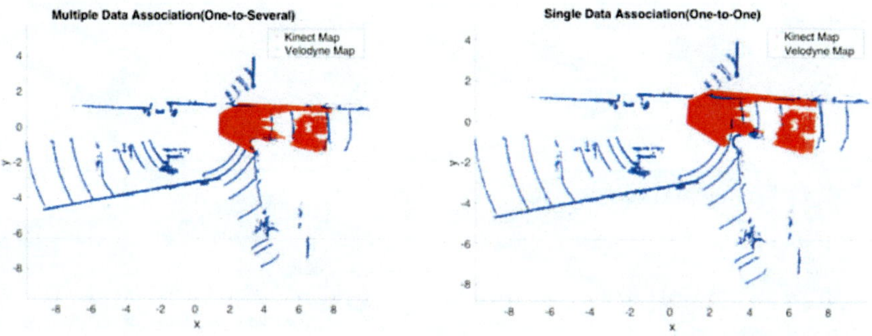

(a) At t=0s, velodyne map is very sparse. Left image shows the sucessful registration of multiple probabilisitc data association (average associated voxels=3.57), while right shows the large offset between two maps by applying one-to-one correspondence.

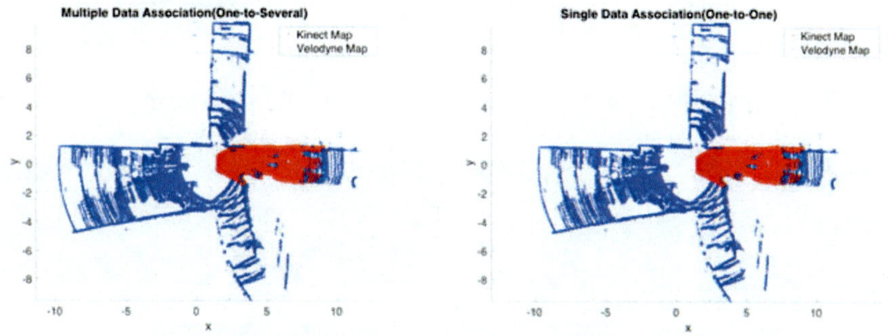

(b) At t=20s, velodyne map gets denser. Both algorithms produce satisfying algorithm. It can be seen that the density of velodyne map is increasing. Thus, the average associated voxels of multiple data association strategy deacrease to 2.15).

Fig. 5.8 The experiment of heterogeneous map matching with different data association strategies

local maps.), the overlapped area between the two maps combines the advantages of the two sensors, which produces a much more detailed and accurate fused map.

5.5.3 Mixed Environment Ground Floor

The mixed environment depicted in Fig. 5.7b is at the ground floor of a building in the university, where the indoor environment is the lobby, and the outdoor environment is the affiliated car park. Two robots were tele-operated to explore the environment with different sensors (Asus Xtion PRO and Velodyne VLP-16).

Figure 5.10 shows the fusion of maps on local level and global level. Robot 1 generates map in an indoor environment, where the Asus Xtion PRO captured most of the information in the lobby. As highlighted in Fig. 5.10 (Top: individual map

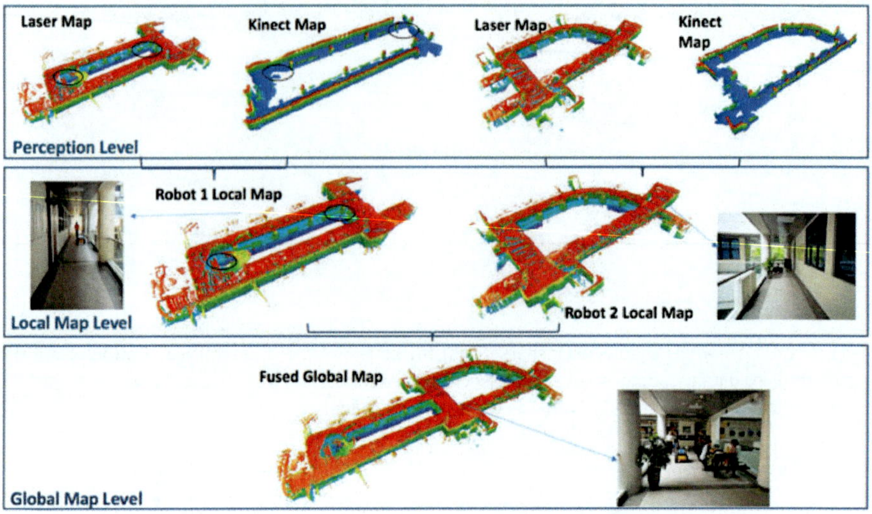

Fig. 5.9 A demonstration of multi-level map fusion result in indoor experiment

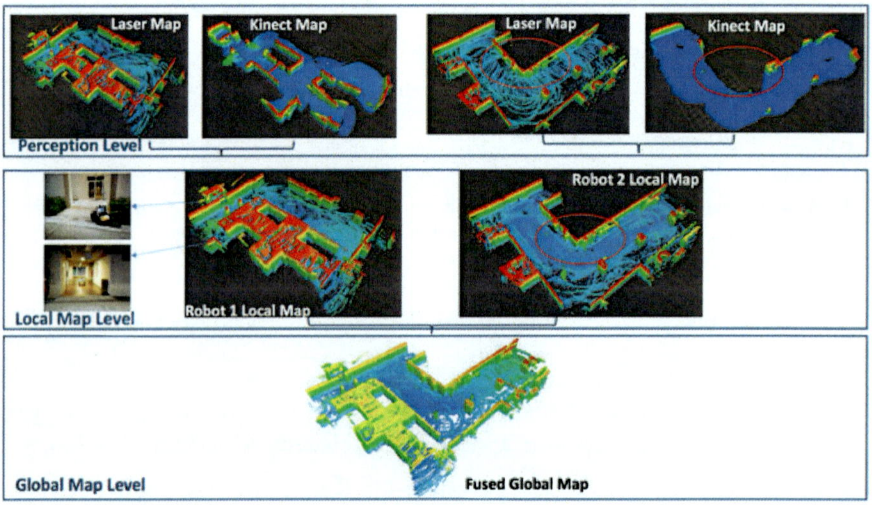

Fig. 5.10 A demonstration of multi-level map fusion result in mix experiment

produced by heterogeneous sensors. Middle: the locally fused map by two different robots. Bottom: the global map by fusing two local maps.), the Kinect map of robot 2 missed the wall due to the limited perception range of the sensor. The velodyne map of robot 2 produces a sparse map, especially for the ground, which is due to the sensor limitation. As underlined with circles, the algorithm combines the map information from two heterogeneous maps to generate an enhanced and more consistent global map.

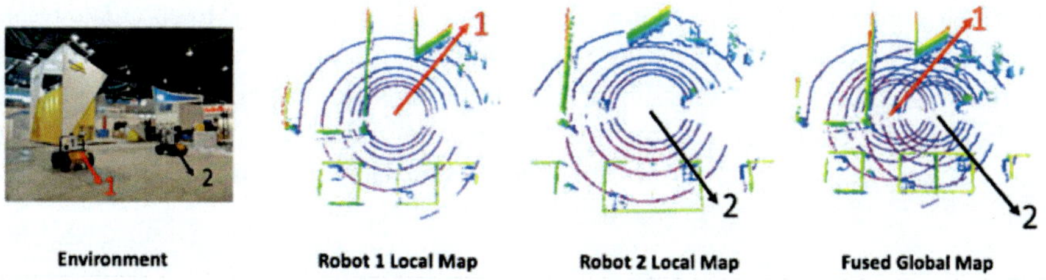

(a) t=0s, the inital position of the two robots. Two sparse locals maps are successfully fused into a global map.

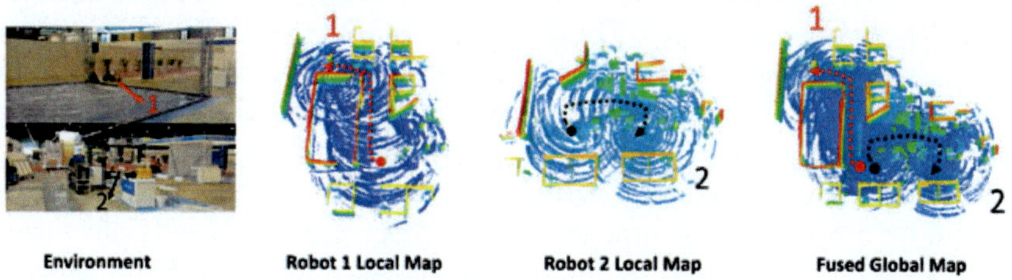

(b) t=76s, half of the mission. The local maps become denser, and global map are fused.

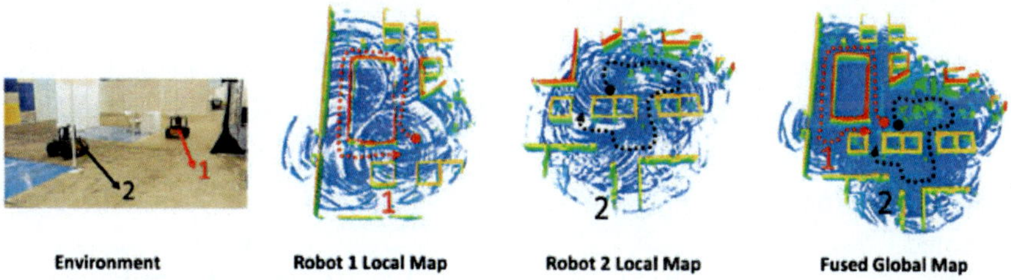

(c) t=145s, the end of the mission. The two robots back to the start position and the global map of the full environment is presented.

Fig. 5.11 A demonstration of the collaborative mapping system in real-time at exhibition center

5.5.4 Changi Exhibition Center

The Changi Center depicted is at a convention center during the Asia Unmanned System Exhibition in Singapore, where the red arrow and trajectory is robot **1**, the black represents robot **2**. Two robots were utilized with the same configuration as the experiment in the mixed ground floor environment. The robots start their mapping from a common neighborhood, and explore the environment by traversing two different paths. It differs from previous two experiments since it also equips Velodyne 3D Lidar. As shown in Fig. 5.11, the fusion starts from very sparse Velodyne maps to denser Velodyne maps, where the fusion process requires the data association to dynamically adjust the numbers of voxels to be associated. In the initial position, a more through and extensive correspondences will be established to calculate relative transformation. Then, the average numbers of voxels to be associated is decreased as the maps grow denser. Finally, the two local maps are fused into a more consistent and enhanced global map. More specifically, the global map filled the gaps in the partial maps of the robots occurred due to loss of data and sensor limitations.

5.5.5 Unstructured Environment Environment

Experiments conducted in the forest environment is presented in this part. Each robot was equipped with a Velodyne VLP-16 Lidar. The forest is in the university with full 3D environment that contains trees and slope. It differs from previous three experiments since it is the fully unstructured environment. The two robots started from nearby place, while robot 1 moved straight forward and robot 2 climbed to a small hill and turned back to the start position. As presented in Fig. 5.12, the map produced by robot 1 is semi-structured, while robot 2 produces fully unstructured forest map. The partial maps are fused into a consistent global map. The algorithm combines the map information from partial maps to generate an enhanced and more consistent global map.

5.5.6 Analysis of Experiment Results

In this subsection, the quantitative results in respect of accuracy and efficiency are presented.

Table 5.2 summarizes the RMS error for all experiments. It can be seem The trend is clear that the smallest errors in Table 5.2 exhibit the highest performance of transformation estimation. During the collaborative mapping process, the RMS errors for all experiments are decreasing. All the algorithms witness a large offset at the beginning of the experiments, which is caused by the low density of the map. As the maps became denser, more correspondences between the voxels can be established and the errors will decrease. In can be also found that the RMS errors in Changi Center and Unstructured Environment are higher compared with Indoor Environment and Mixed Environment. The reason is that Changi Center and Unstructured Environment are highly unstructured environment, while Indoor Environment and Mixed Environment are more structured.

Our methods outperforms OICP in all of the cases. In Indoor Environment and Mixed Environment experiments, two robots were equipped with heterogeneous sensors, our algorithm dynamically adjust the data association strategy in the matching (See Fig. 5.13). As a consequence of this strategy, the RMS error of our algorithm improved significantly of 10–20% compared to the one-to-one strategy OICP in the two scenarios. In unstructured Changi Center and Unstructured Environment, only Velodyne 3D Lidar was equipped due to the sensor limitation of Asus Xtion Camera.

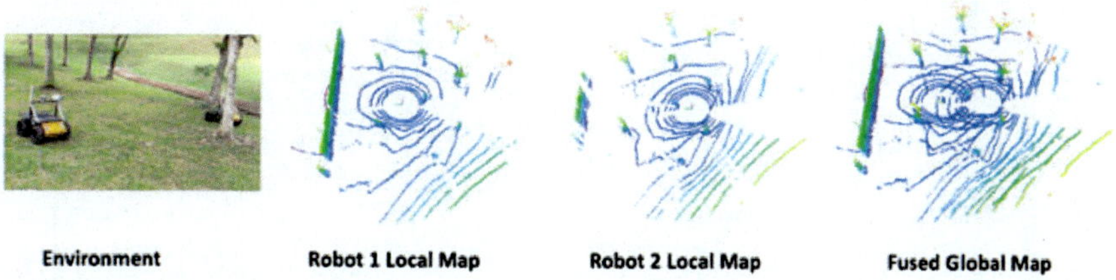

(a) t=0s, the inital position of the two robots. Two sparse locals maps are successfully fused into a global map.

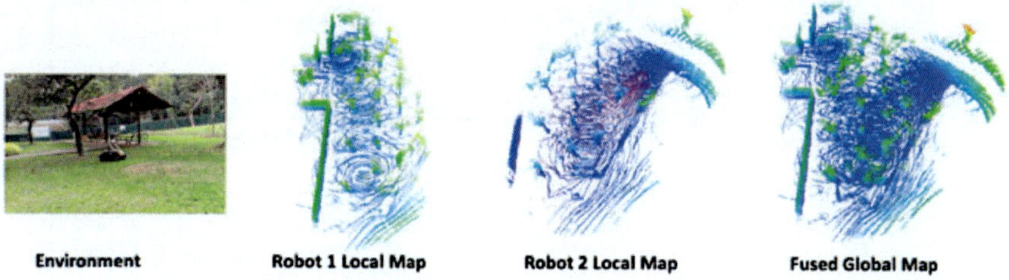

(b) t=120s, half of the mission. The local maps become denser, and global map are fused.

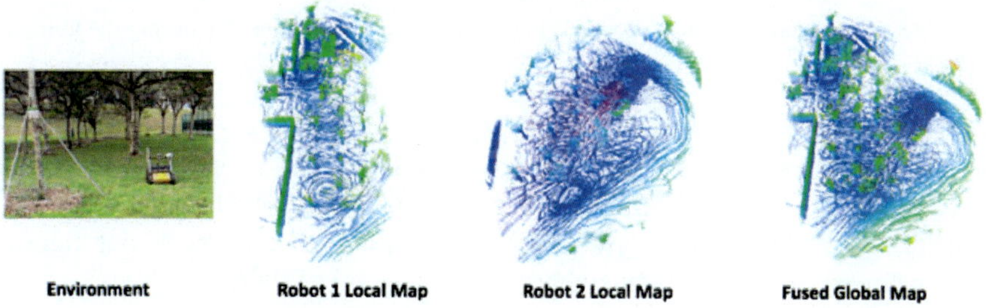

(c) t=215s, the end of the mission. The global map of the full environment is presented.

Fig. 5.12 A demonstration of the collaborative mapping system in real-time at forest environment

Table 5.2 The root mean error in all experiments. (The best performance is denoted in bold.)

Datasets	Beginning		Middle		Final	
	OICP	Ours	OICP	Ours	OICP	Ours
Indoor Environment	0.3414	0.2479	0.2592	0.2357	0.2675	0.2890
Mixed Environment	0.3140	0.2791	0.3078	0.2563	0.2818	0.2542
Changi Center	0.4405	0.4284	0.4054	0.3998	0.3653	0.3550
Unstructured Environment	0.4393	0.4301	0.3561	0.3374	0.3430	0.3237

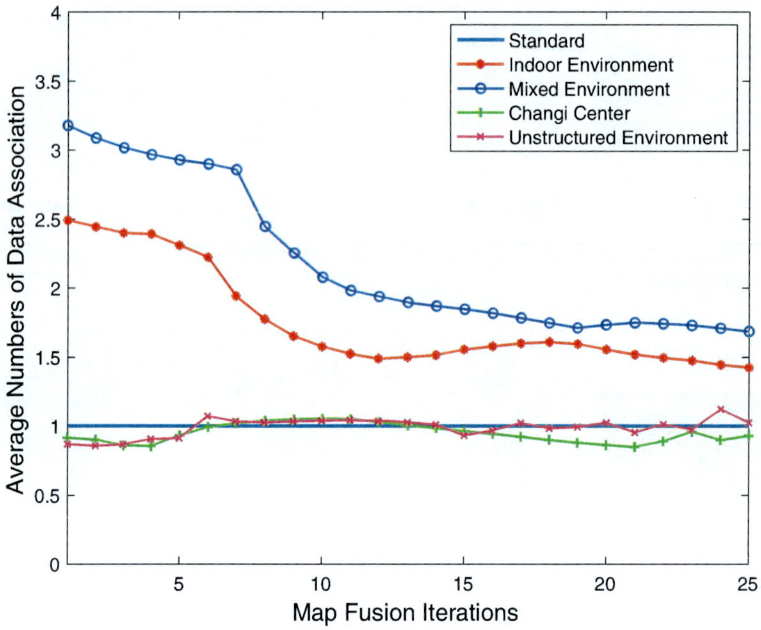

Fig. 5.13 The average numbers of voxels to be associated in our algorithm during the four experiments

In these two experiments, the numbers of voxels to be associated are close to the one-to-one (See Fig. 5.13). Both OICP and our algorithm achieve comparable convergence results and our algorithm still outperforms OICP. Overall, our algorithm demonstrates superior or comparable performance over OICP.

Additionally to the evaluation of the error metrics, computation times are an important factor in the real-time robotics applications. Figure 5.14 summarizes the computational times of the four experiments when executed on an Intel Core i7-6700HQ CPU @ 2.60GHz robot onboard computer. The efficiency is reported per full calculation time of the map fusion process of every running. It can be found that all experiments were running in semi-real time. The computational time for Unstructured Environment is significantly higher than other scenarios due to the large scale and unstructured environment. For collaborative mapping missions, we are considering the strategy of local maps from individual robots being fused sporadically. Considering connectivity/bandwidth and computational constraints, transferring map data only when local maps are significantly updated is preferred over transferring every simple map update. Therefore, the time interval between two such fusion processes is usually in the order of minutes, where the map fusion process does not operate at high frequency. Hence execution time of several seconds (i.e. semi-real time) for map fusion is justified.

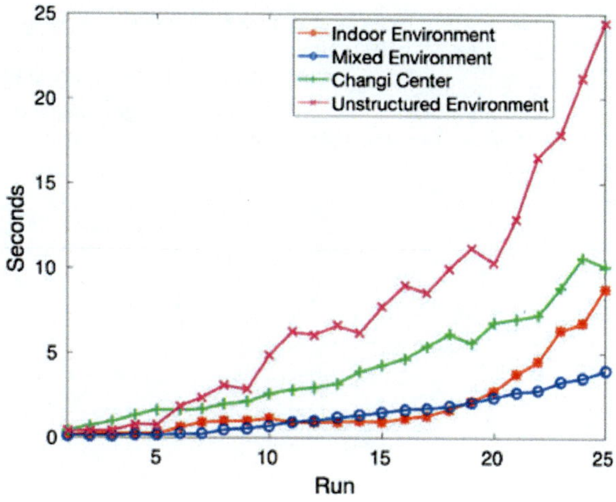

Fig. 5.14 The processing time during the four experiments

5.6 Conclusions

In this chapter, a general probabilistic framework for multi-robot 3D mapping using heterogeneous sensors is proposed, which is independent of sensor types and SLAM algorithms. The framework provides theoretical basis for computing the relative transformations (map matching) and merging probabilistic map information (map merging), which enables the team of robots to work with different sensor types on different platforms. Since maps produced by heterogeneous sensors poss diverse physical properties, a probabilistic map matching algorithm is proposed that automatically decides the numbers of voxels to be associated. To distribute the maps among robots, a time-sequential map merging framework that makes fusing maps from distributed multi-robot system efficiently, where uncertainty embedded in the map and caused by the relative transformation are propagated and integrated. The proposed approach is evaluated in real-time experiments in various environments with heterogeneous sensors. The results verify the accuracy, efficiency and generality for collaborative mapping missions.

References

1. Yan Z, Jouandeau N, Cherif AA (2013) A survey and analysis of multi-robot coordination. Int J Adv Robot Syst 10(12):399
2. Saeedi S, Paull L, Trentini M, Seto M, Li H (2014) Group mapping: a topological approach to map merging for multiple robots. IEEE Robot Autom Mag 21(2):60–72
3. Jessup J, Givigi SN, Beaulieu A (2015) Robust and efficient multirobot 3-d mapping merging with octree-based occupancy grids. IEEE Syst J 11(3):1723–1732
4. Yue Y, Senarathne PN, Yang C, Zhang J, Wen M, Wang D (2018) Hierarchical probabilistic fusion framework for matching and merging of 3-d occupancy maps. IEEE Sens J 18(21):8933–8949

5. Bonanni TM, Della Corte B, Grisetti G (2017) 3-d map merging on pose graphs. IEEE Robot Autom Lett 2(2):1031–1038
6. Yue Y, Wang D, Senarathne P, Moratuwage D (2016) A hybrid probabilistic and point set registration approach for fusion of 3d occupancy grid maps. In: 2016 IEEE international conference on systems, man, and cybernetics (SMC), IEEE, pp 001975–001980

Chapter 6
All-Weather Collaborative Mapping with Dynamic Objects

Enabling long-term operation during day and night for collaborative robots requires a comprehensive understanding of the unstructured environment. Besides, in the dynamic environment, robots must be able to recognize dynamic objects and collaboratively build a global map. This chapter proposes a novel approach for dynamic collaborative mapping based on multimodal environmental perception. For each mission, robots first apply heterogeneous sensor fusion model to detect humans and separate them to acquire static observations. Then, the collaborative mapping is performed to estimate the relative position between robots and local 3D maps are integrated into a globally consistent 3D map. The experiment is conducted in the day and night rainforest with moving people. The results show the accuracy, robustness, and versatility in 3D map fusion missions.

6.1 Introduction

In Chap. 5, the problem of collaborative localization and mapping in unstructured environment is discussed. However, the environment is assumed to be static and dynamic objects such as pedestrians are ignored. To improve the intelligence of robots, and make their behavior more targeted according to missions. The movement behavior of dynamic humans in the environment need to be analyzed so that the mission-critical target can be discovered. Till now, teams of robots that can adapt to unstructured and dynamic environments remain a challenging problem. In the genuinely GPS-denied and dynamic unstructured environment, collaborative robots needs to achieve precise localization and collaborative dynamic mapping in the cluttered unknown scenario [1, 2]. In this chapter, the collaborative dynamic mapping is achieved based on multimodal environmental perception.

Y. Yue and D. Wang, *Collaborative Perception, Localization and Mapping for Autonomous Systems*, Springer Tracts in Autonomous Systems 2,
https://doi.org/10.1007/978-981-15-8860-0_6

Collaborative robots 3D environmental mapping is generally called collaborative mapping, where attempts have been made from different aspects to address it, such as estimating the relative transformation between the robots [3], accounting for the uncertainty of the relative localization [4], integrating the information from neighboring robots [5], updating maps in the global perspective [6], and many more. However, the methods mentioned above assume that the world is static, dynamic objects such as pedestrians are ignored. In addition, they can't provide day and night operation ability in unstructured environment.

Collaborative robots are expected to operate in increasingly complex environments, such as mapping in the presence of pedestrians at day and night. For example, a visual camera does not guarantee high quality at night, while a thermal camera can accurately capture human motion, providing the basis for robot long-term operation [7]. This fusion scheme takes advantage of the characteristics of different modalities to enhance the data authenticity and availability [8]. Among multiple robots, data fusion can be grouped into three different types, raw sensor data [9], volumetric maps [10] and topological maps [11]. Although raw sensor data offers abundant description of the real circumstance, they are difficult to process and transmit with limited communication bandwidth and computing power. Therefore, raw sensor data is not suitable to be shared for collaborative mapping tasks. Metric grid maps such as 3D

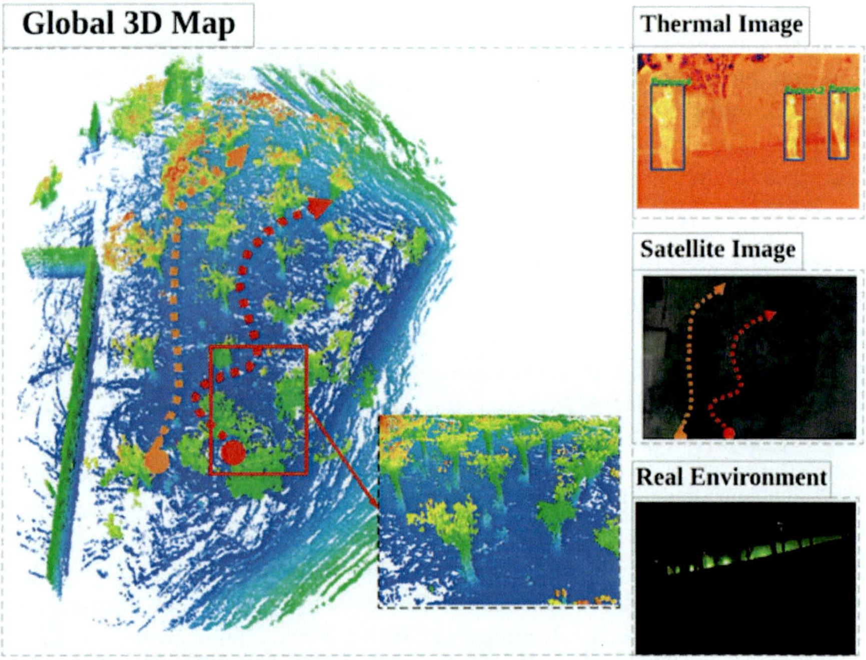

Fig. 6.1 A demonstration of the simultaneous collaborative mapping in rainforest at night. Red trajectory is robot 1, and orange trajectory is robot 2. Left: collaboratively generated 3D global map. Right: an overview of the environment setting and robot perception result

volumetric maps [12] models the environment into probabilistic form with sufficient detail. Combined with the compressed size of the map, it is chosen as the choice of map type.

This work is motivated by the fact that an approach for collaborative mapping in day and night dynamic environment is not studied in depth. This chapter comprehensively models and implements collaborative dynamic mapping in day and night unstructured environments. A general collaborative dynamic mapping scheme is proposed that provides flexibility for different types of sensors and SLAM methods. By detecting and filtering out dynamic people in the environment, the robot can achieve more accurate relative positioning and global mapping. The approach has been evaluated in various unstructured dynamic environments at day and night, which exhibits higher accuracy, efficiency and robustness compared to existing methods (see Fig. 6.1).

The rest of the chapter is organized as follows. Section 6.2 gives an overview of the proposed framework. The multimodal environmental perception is detailed in Sect. 6.3. Section 6.4 explains the theoretical basis for collaborative mapping. Section 6.5 shows the experimental procedures and results. Section 6.6 concludes the chapter with a discussion on future work.

6.2 Framework of Collaborative Dynamic Mapping

In this section, the system architecture for collaborative dynamic mapping is presented, where the problem is formulated to provide a theoretical basis for the collaborative framework on the single robot and collaborative robots level.

The objective of this chapter is to develop a common framework for collaborative mapping that will estimate the global map in dynamic environments. The overview of the system architecture is depicted in Fig. 6.2, where the framework consists of three modules: *multimodal environmental perception, single robot mapping, collaborative dynamic mapping*. Since the proposed collaborative robots operate in distributed setup, the organization of the chapter follows the order from a single robot to multiple robots.

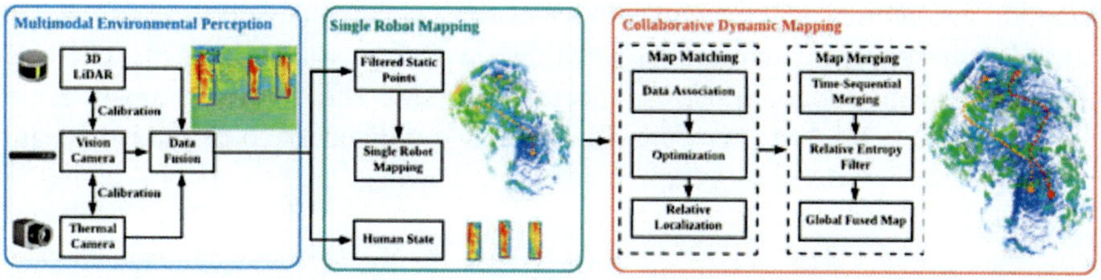

Fig. 6.2 The diagram of the whole system. Multimodal environment perception is detailed in Sect. 6.3. Single robot mapping and collaborative dynamic mapping are presented in Sect. 6.4

Table 6.1 The main notations used throughout the paper

Symbol	Description
	Single Robot Level:
r	A single robot is denoted as r
$z_t^{(r)}$	The sensor observation of robot r at time t
$u_t^{(r)}$	Control commands applied to robot r at time t
$x_t^{(r)}$	Pose of robot r at time t
$m_t^{(r)}$	Local map generated by robot r at time t
	Collaborative Robots Level:
$\mathcal{R}$	Set of all robots $\mathcal{R} = \{r\}_{1:R}$, R is number of robots
$\mathcal{R}_r$	Set of neighboring robots $\mathcal{R}_r = \{r_n\}_{1:R_n}$ in communication with r, where the number is R_n
M_t	The global map generated at time t
T_{r,r_n}	Relative transformation between robot r and r_n
$T_{r,\mathcal{R}_r}$	Set of relative transformation between robot r and all neighboring robots $\mathcal{R}_r$

For single robot level, the heterogeneous sensors carried by each robot are calibrated and integrated. The object detection and tracking algorithms are executed to separate the static point cloud and human stream in two parallel processes. Then, each robot performs static environmental SLAM given the filtered static point cloud as input.

For collaborative robots level, each robot communicates with neighboring robots to share local 3D maps estimated by single robot. The collaborative robots perform collaborative dynamic mapping, estimating relative position between all robots and global 3D map. The main notational symbols are defined in Table 6.1.

6.3 Multimodal Environmental Perception

This section aims to develop the multimodal environment perception algorithm. To provide a comprehensive understanding of the complicated surroundings, heterogeneous sensors are calibrated. Meanwhile, involving dynamic humans introduced difficulty for the robots to perform accurate SLAM. The dynamic humans are filtered out to obtain a static 3D point cloud, which is utilized for collaborative dynamic mapping.

6.3.1 Heterogeneous Sensors Calibration

Heterogeneous sensors provide complementary characteristics, and it is essential for robots to achieve a robust and long-term perception. Specifically, 3D LiDARs can accurately record a 360 degree view of the environment with accurate depth information, which is vital for robust large-scale localization and mapping in unstructured environments. However, due to the particularity of the sensor, 3D LiDAR can only provide sparse geometric information, which makes it challenging to detect objects. Visual and thermal cameras can detect and track dynamic objects with high accuracy in day and night. To deeply fuse these sensors, a precise geometric relationship (rotation and translation) between the sensor frames needs to be established. To obtain the relationship, accurate calibration is indispensable.

Initially, the geometric relationship between multiple sensors mounted on a robot is unknown. To obtain the 6 DOF extrinsic parameters among three sensors (a 3D LiDAR, a visual and a thermal camera), a two-step method we have proposed is utilized [13]. First, the extrinsic parameter between a visual camera and a thermal camera $\mathbf{T}_c^t$ is obtained via stereo camera calibration. Second, the extrinsic parameter between a visual camera and a 3D LiDAR $\mathbf{T}_l^c$ is obtained at last, $\mathbf{T}_l^t$ can be calculated by multiplying $\mathbf{T}_c^t$ and $\mathbf{T}_l^c$. Once the transformation parameters are obtained, the three sensors can be fused together for further applications.

6.3.2 Separation of Static and Dynamic Observations

In a dynamic environment, suppose the sensor observation is denoted as $z_t^{(r)} = \{s_t^{(r)}, b_t^{(r)}\}$, where $s_t^{(r)}$ is static world observation, $b_t^{(r)}$ is dynamic human observation. To separate the dynamic objects $b_t^{(r)}$ and the static environments $s_t^{(r)}$, the process described in Fig. 6.3 is perform. First, the dynamic humans are detected. Directly detecting objects in a sparse point cloud remains to be a challenging issue, in our approach, objects are detected from visual or thermal images. In our approach, YOLOv3 [14] is chosen. By projecting the 3D laser points to the image frame, we can also calculate the real distance of the objects in the environment. Then, a static point cloud is generated by filtering out the dynamic humans from the raw 3D point cloud.

6.4 Distributed Collaborative Dynamic Mapping

In this section, a distributed collaborative dynamic mapping framework is presented. Single robot SLAM is performed first to produce local map for each robot, which is the input for collaborative map fusion. Then, the framework jointly estimates

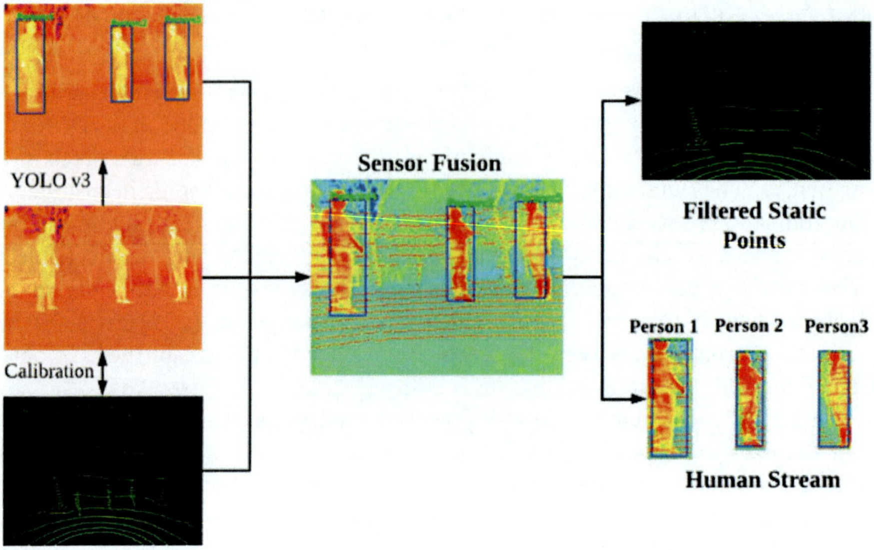

Fig. 6.3 The overview of robot environmental perception results. Multimodal information has been integrated to provide a comprehensive understanding of the environment

the posterior of computing the relative transformations between robots and merging probabilistic map information.

6.4.1 Single Robot Level Definition

For each robot r, the objective is to estimate its local map $m_t^{(r)}$, trajectory $x_{1:t}^{(r)}$, human state $g_{1:t}^{(r)}$ given sensor observations $z_{1:t}^{(r)}$ and control sequences $u_{1:t}^{(r)}$.

$$p(m_t^{(r)}, x_{1:t}^{(r)}, g_{1:t}^{(r)} | z_{1:t}^{(r)}, u_{1:t}^{(r)}) \tag{6.1}$$

The novel formulation of Eq. (6.1) enables the static and dynamic perception to be performed at the same time, which acts as the theoretical basis for single robot scene understanding.

In the framework, multimodal environmental perception serves as the input for single robot level as well as the whole framework. As defined in Table 6.1, raw sensor observation $z_{1:t}^{(r)}$ is composed of static world observation $s_{1:t}^{(r)}$ and dynamic object observation $b_{1:t}^{(r)}$. Object detection algorithm is applied to detect dynamic object $b_{1:t}^{(r)}$ (human in this chapter) from $z_{1:t}^{(r)}$. Then, $s_{1:t}^{(r)}$ is acquired by removing the dynamic objects $b_{1:t}^{(r)}$.

Hence, the single robot SLAM problem Eq. (6.1) can be factorized as a product of single robot Static SLAM problem and human state estimation problem conditioned on the robot pose. Then, Eq. (6.1) is factorized into Eq. (6.2).

$$p(m_t^{(r)}, x_{1:t}^{(r)}, g_{1:t}^{(r)} | z_{1:t}^{(r)}, u_{1:t}^{(r)}) = \underbrace{p(m_t^{(r)}, x_{1:t}^{(r)} | s_{1:t}^{(r)}, u_{1:t}^{(r)})}_{\text{Static SLAM}} \cdot \underbrace{p(g_{1:t}^{(r)} | b_{1:t}^{(r)}, x_{1:t}^{(r)})}_{\text{Dynamic Estimation}} \tag{6.2}$$

In Eq. (6.2), the input to two sub-problems are static observation $s_{1:t}^{(r)}$ and control sequence $u_{1:t}^{(r)}$, respectively. As a result, the generated $m_t^{(r)}$ is the static environment map. This will increase the accuracy and robustness of localization and mapping in dynamic environment. Then, human state $g_{1:t}^{(r)}$ is estimated given the robot pose $x_{1:t}^{(r)}$ and dynamic object observation $b_{1:t}^{(r)}$.

A single SLAM problem is to estimate the trajectory of the robot $x_{1:t}^{(r)}$ at each time step and the posterior over the map $m_t^{(r)}$ given static world observation $s_{1:t}^{(r)}$ and control sequences $u_{1:t}^{(r)}$ (see Eq. (6.3)). The single robot SLAM problem has been well studied and many efficient approaches have been developed. Figure 6.4 shows the mapping results using different 3D points, and the red box highlights their differences. It can be seen that by filtering out the moving people, the accuracy and consistency of the map has been greatly improved.

$$p(m_t^{(r)}, x_{1:t}^{(r)} | s_{1:t}^{(r)}, u_{1:t}^{(r)}) = \underbrace{p(x_{1:t}^{(r)} | s_{1:t}^{(r)}, u_{1:t}^{(r)})}_{\text{Localization}} \cdot \underbrace{p(m_t^{(r)} | x_{1:t}^{(r)}, s_{1:t}^{(r)})}_{\text{Mapping}} \tag{6.3}$$

6.4.2 Collaborative Robots Level Definition

The objective of distributed collaborative dynamic mapping is to estimate fused global map M_t, set of relative transformation $T_{r,\mathcal{R}_r}$ under a fully distributed network, given local maps $m_t^{(r,\mathcal{R}_r)}$ from neighboring robots $\mathcal{R}_r$.

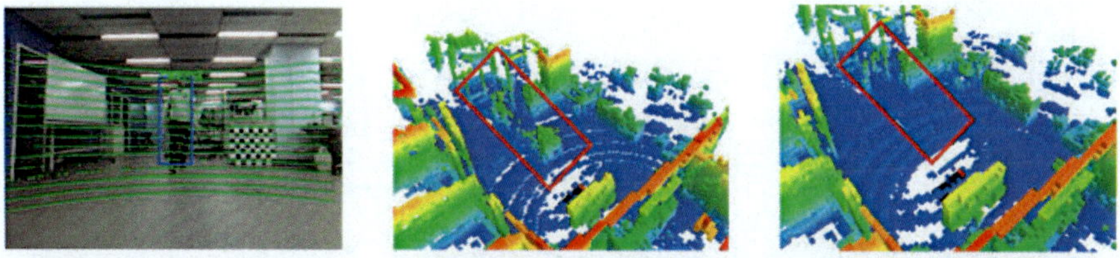

(a) Raw sensor data **(b) 3D Mapping using raw laser points** **(c) 3D Mapping using filtered laser points**

Fig. 6.4 The difference of using raw laser points and filtered laser points for single robot SLAM. The inconsistent 'wall' in (**b**) is caused by the moving person

$$p(M_t, T_{r,\mathcal{R}_r} | m_t^{(r,\mathcal{R}_r)}) \tag{6.4}$$

The objective of collaborative dynamic mapping is to estimate a set of relative transformations $T_{r,\mathcal{R}_r}$ and generate the global map M_t on the basis of all available partial maps $m_t^{(r,\mathcal{R}_r)}$. The set of relative transformation is denoted as $T_{r,\mathcal{R}_r} = \{T_{r,r_n}, r_n \in \mathcal{R}_r\}$, where T_{r,r_n} is the transformations between robot r and r_n.

When robot r receives the partial maps $m_t^{(\mathcal{R}_r)}$ from neighboring robots $\mathcal{R}_r$, the probabilistic map fusion algorithm described below is performed to generate global map M_t. Each robot r maintains its own global map, which increases the robustness of the mapping mission to unexpected robot breakdowns.

The joint distribution of collaborative mapping of Eq. (6.4) can be factorized as the product of two parts (see Eq. (6.5)), the relative transformation posterior and the global map posterior, which is given as follows.

$$p(M_t, T_{r,\mathcal{R}_r} | m_t^{(r)}, m_t^{(\mathcal{R}_r)}) = \underbrace{p(T_{r,\mathcal{R}_r} | m_t^{(r)}, m_t^{(\mathcal{R}_r)})}_{\text{Map Matching}} \cdot \underbrace{p(M_t | T_{r,\mathcal{R}_r}, m_t^{(r)}, m_t^{(\mathcal{R}_r)})}_{\text{Map Merging}} \tag{6.5}$$

This decomposes the collaborative map fusion problem into two consecutive subproblems, namely *map matching* and *map merging*, as demonstrated in Fig. 6.5. It places no constraint on the type of sensor data or SLAM algorithm used. Regardless of the SLAM algorithm and sensor type used, map matching works if accurate partial map is provided. In addition, it benefits from utilizing a relative entropy filter based map merging algorithm.

To estimate $T_{r,\mathcal{R}_r}$, we assume an initial transformation is available. Firstly, data association needs to be estimated , which is addressed by performing the nearest neighbor search of Mahalanobis distance. Then, the optimization algorithm is performed to minimize the error metric. This process will be performed iteratively until it achieves convergence. The map merging is performed in serial since the local maps are received in sequence. A relative entropy filter [15] is applied to measure the cor-

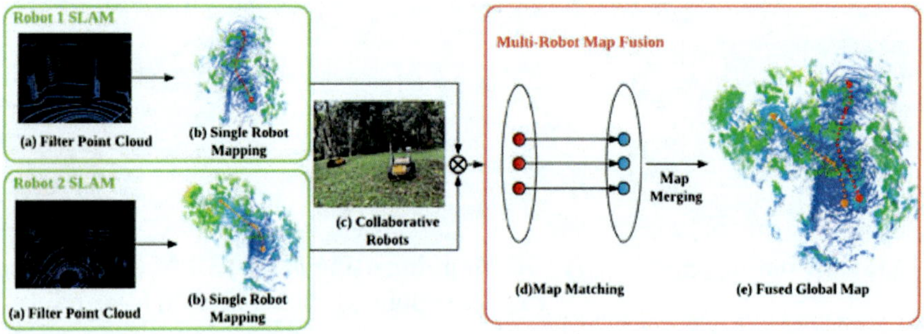

Fig. 6.5 The overview of collaborative dynamic mapping. **a**, **b** Single robot SLAM is perform to generate the local 3D map. **c** Distributed map sharing. **d** Map matching for relative position estimation. **e** Time sequential map merging

responding pairs, where the fusion leads to a higher entropy will be rejected. Then, a global fused map is generated by combing the information from partial maps.

6.5 Experiments

Extensive unstructured scenarios are investigated at Nanyang Technological University, including Daytime Unstructured (Sect. 6.5.2) and Night-time Unstructured (Sect. 6.5.3). Both qualitative and quantitative results are presented. The former is obtained by showing the results of each stage in various situations. Then, the latter is expressed by analyzing the performance.

6.5.1 Experiments Overview

6.5.1.1 Experiment Setup

Experiments are conducted by operating two robots in two different scenarios in Nanyang Technological University (see Table 5.1). All algorithms are executed on the ROS platform for multimodal environmental perception and collaborative dynamic mapping. Two robot platforms (Husky) are equipped with the Intel Core i7-6700HQ CPU @ 2.60GHz CPU and the Nvidia GeForce RTX 2060 @ 6GB RAM GPU. Each robot is equipped with a long-range 3D LiDAR Velodyne VLP16, a ZED stereo camera (resolution: 672×376) and an Optris PI450 thermal camera (resolution: 382×288). The communication between robots is established by long range Wi-Fi with limited bandwidth.

Due to the limited computational resources onboard, all incoming data are running at 10 Hz. In all cases, YOLOv3 [14] is used for moving people detection. Then, the static and dynamic observations are separated. LOAM [16] is applied for single robot pose estimation, and Octomap [12] is utilized for single robot mapping at a resolution of 0.1m. The experiment environment settings are:

- **Daytime Unstructured Environment**: two Husky robots equipped with 3D LiDAR and visual camera in a dense rainforest.
- **Night-time Unstructured Environment**: one Husky robots equipped with visual camera and 3D LiDAR, while another one equipped with thermal camera and 3D LiDAR in a forest at night.

Figure 6.6 summarize the settings and the main characteristics of each experiment.

(a) Daytime Unstructured (b) Night-time Unstructured (c) Robot Platform

Fig. 6.6 Experimental environments and setups

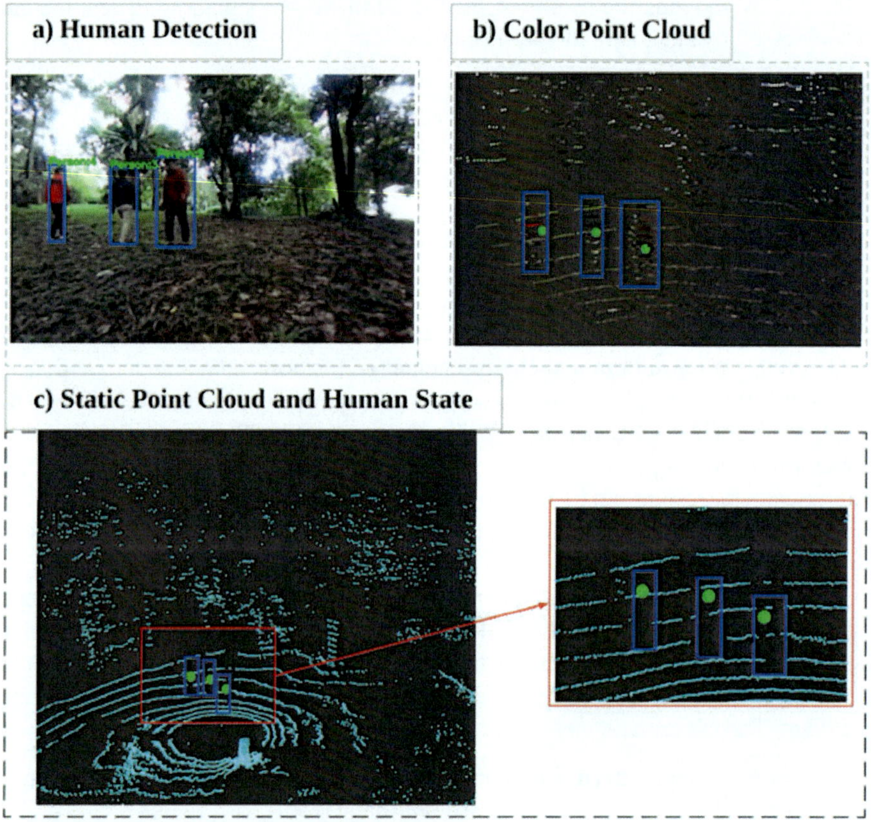

Fig. 6.7 The results of multimodal environmental perception in daytime rainforest

6.5.1.2 Comparison Baseline

The collaborative dynamic mapping is compared against [17]. Since the code of [17] is not available, we implemented it based on the standard Point Cloud Library [18]. The chapter compares the metric Mean Squared Error (MSE) for collaborative map-

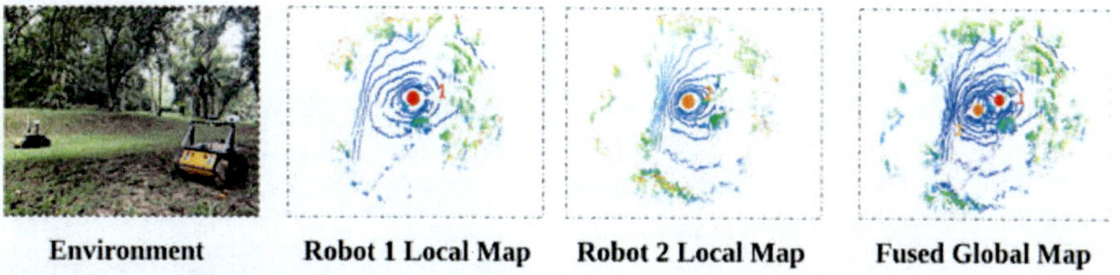

(a) The inital position of the two robots. Two sparse locals maps are successfully fused into a global map.

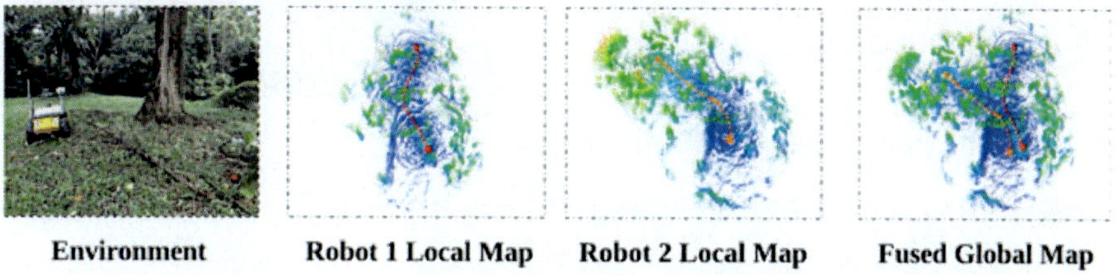

(b) Half of the mission. The local maps become denser, and global map are fused.

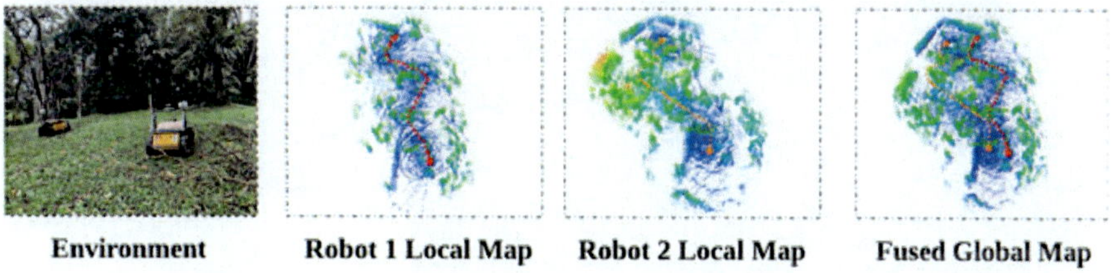

(c) End of the mission. The global map of the full environment is presented.

Fig. 6.8 The result of collaborative static mapping in day rainforest environment. The proposed method can reconstruct detailed 3D global map

ping accuracy, which is the mean squared Euclidean distance between the matched map voxels.

6.5.2 Daytime Unstructured Environment

Figure 6.7a shows the detected people with labelled bounding box. By re-projecting the 2D image to the 3D point cloud, we can visualize the 3D color point cloud of the people in Fig. 6.7b. Then, we get filtered static point cloud in Fig. 6.7c, where each moving person is represented with a labelled bounding box. The magnified image shows the existence of three detected people in the point cloud.

The two robots started moving from a nearby place, while robot **1** (red trajectory) turned left and went uphill and robot **2** (orange trajectory) turned right and went downhill. We validate the results of the collaborative mapping results in Fig. 6.8. The two local maps are successfully matched and merged with high accuracy.

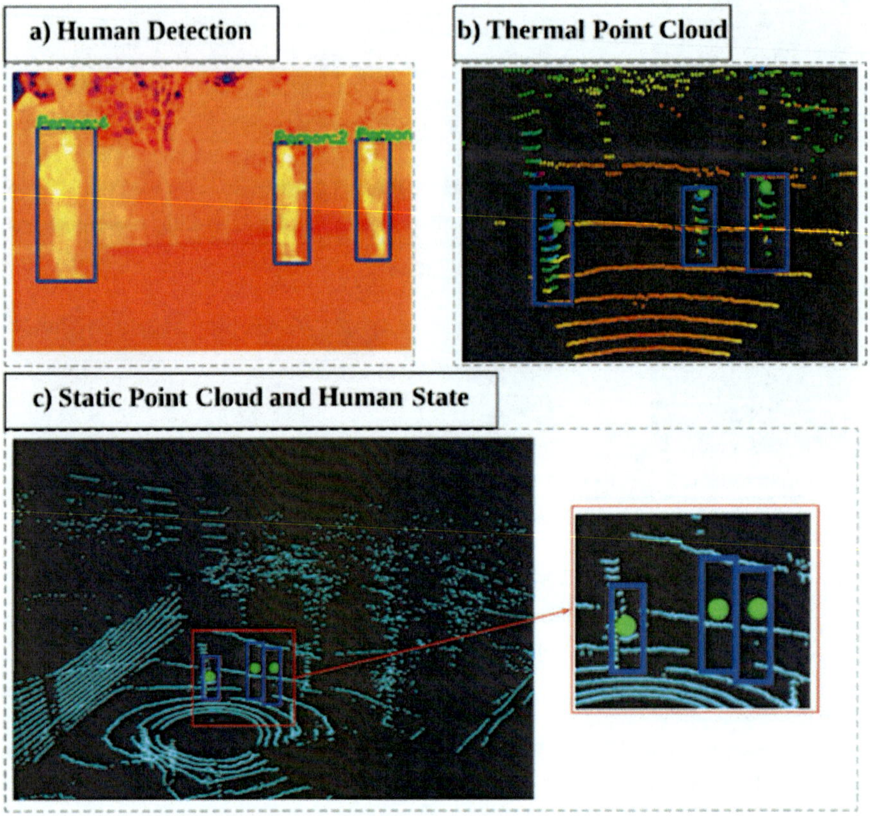

Fig. 6.9 The results of multimodal environmental perception in night forest

6.5.3 Night-Time Unstructured Environment

Different from the previous daytime forest, this experiment is conducted on a dark night. The thermal images are input into YOLOv3 for human detection. Figure 6.9a shows detected people with labels. Figure 6.9b is acquired by projecting thermal image to 3D laser points. The different colors in the 3D point cloud represent different temperatures, so we call it the thermal point cloud. Finally, Fig. 6.9c presents filtered static point cloud and human states.

The two robots started nearby, while robot **1** (denoted in red) climbed to a hill and robot **2** (denoted in orange) moved straight forward. As the visual image is not visible at night, we show the thermal images instead. As shown in Fig. 6.10, the partial maps are successfully fused into a consistent global map. Although the experiment is conducted at night, the accuracy of the fused global map remains the same high quality as during the day. It shows the robustness of the perception system and the collaborative mapping algorithm, which combines map information to generate an enhanced and more consistent global map at night.

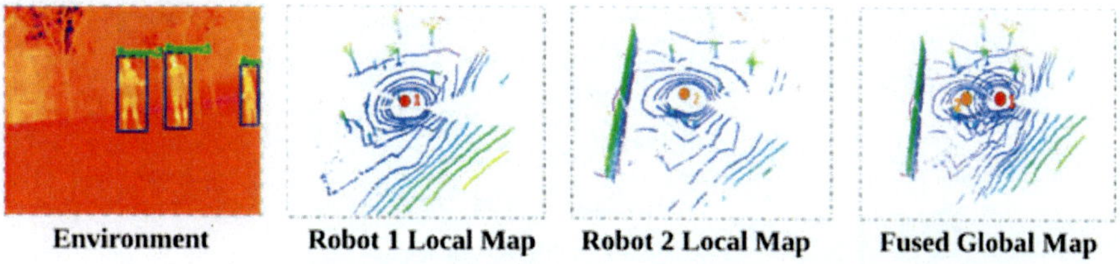

(a) The initial position of the two robots. Two sparse locals maps are successfully fused into a global map.

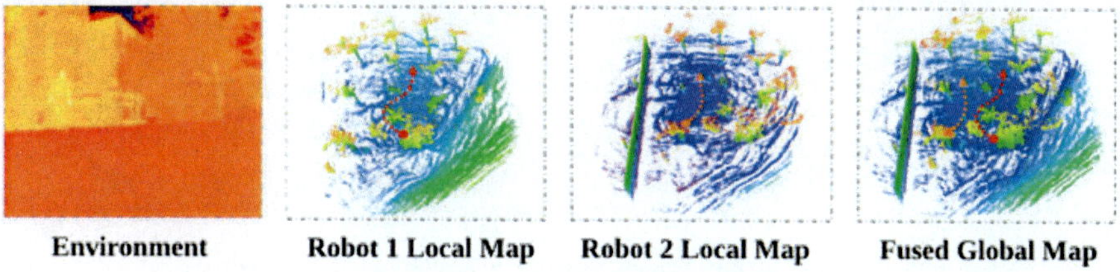

(b) Half of the mission. The local maps become denser, and global map are fused.

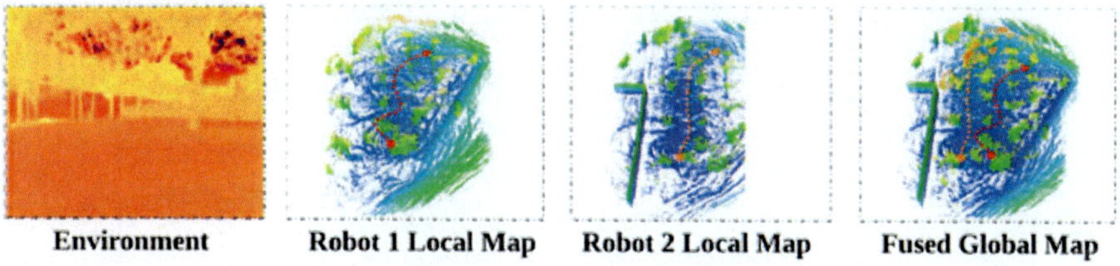

(c) End of the mission. The global map of the full environment is presented.

Fig. 6.10 The results of collaborative static mapping in night forest environment. The thermal images show the night environment, and the proposed algorithm generates high quality map as daytime

Table 6.2 The root mean error in all experiments. (The best performance is denoted in bold.)

Experiments	Beginning		Middle		Final	
	Jessup [17]	Ours	Jessup [17]	Ours	Jessup [17]	Ours
Daytime Unstructured	0.4221	**0.4004**	0.3411	**0.3354**	0.3256	**0.3183**
Night-time Unstructured	0.4247	**0.4173**	0.3171	**0.3095**	0.2980	**0.2908**

6.5.4 *Quantitative Analysis*

Table 6.2 summarizes the MSE error of collaborative mapping for the two experiments. It is measured by the error between the matched common areas of local maps.

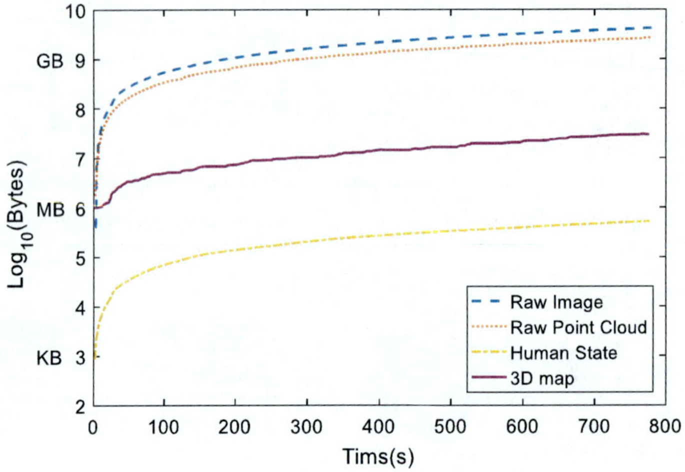

(a) Size of Data in daytime unstructured environment.

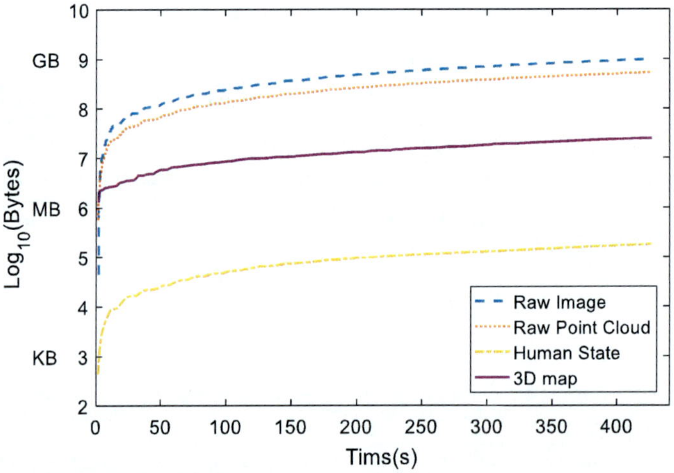

(b) Size of Data in night-time unstructured environment.

Fig. 6.11 The results of cumulative data transmission in all experiments

The smallest errors in Table 6.2 exhibit the highest performance of transformation estimation. Our methods outperform [17] in the two cases. Both the baseline and our algorithms achieve comparable convergence results. Overall, our algorithm demonstrates superior performance over the baseline. Both two algorithms witness a large offset at the beginning of the experiments, which is caused by the low density of the map in the forest environment. As the maps become denser, more correspondences between the voxels can be established and the errors will decrease. As a result, the proposed approach can take the advantage of removing the dynamic objects in the mapping process and utilizing the flexible Expectation-maximization based map matching algorithm to generate more accurate relative transformation estimation.

For real world applications, the size of data need to be transferred between robots is also a key factor. Therefore, the cumulative data size from the beginning to the end of the task is shown in Fig. 6.11. For all experiments, the amount of data for the raw image and point cloud generated by each robot exceeds 1 GB. In contrast, the 3D map generated by each robot is less than 30 MB, and the amount of data of the human state is even less than 1 MB. Therefore, we can find that sharing 3D maps will greatly reduce the data to be transmitted between robots.

6.6 Conclusions

This chapter presented an approach for collaborative dynamic mapping in day and night unstructured environments. We have designed a new framework to provide theoretical formulas and system implementations for collaborative dynamic mapping. Each robot in the group first executed the separation of dynamic and static observations based on heterogeneous sensor fusion and object detection. Explicitly environmental mapping under limited communication bandwidth and computing resources was addressed by developing a collaborative dynamic mapping process. The method was evaluated during day and night in unstructured environment, which exhibited improved precision, efficiency and robustness compared to existing methods.

For hybrid robot teams (UGV and UAV), the vast differences in perspectives present challenges for calculating accurate relative position. It is hoped that the CNN-based method can be used to assign correspondence at the semantic level and improve the accuracy.

References

1. Yue Y, Yang C, Wang Y, Senarathne PCN, Zhang J, Wen M, Wang D (2019) A multilevel fusion system for multirobot 3-d mapping using heterogeneous sensors. IEEE Syst J 14(1):1341–1352
2. Yue Y, Yang C, Zhang J, Wen M, Wu Z, Zhang H, Wang D (2020) Day and night collaborative dynamic mapping in unstructured environment based on multimodal sensors. In: 2020 IEEE international conference on robotics and automation (ICRA)
3. Niedfeldt PC, Speranzon A, Surana A (2015) Distributed map fusion with sporadic updates for large domains. In: 2015 IEEE international conference on robotics and automation, pp 2806–2813
4. Yue Y, Wang D, Sena PGCN, Yang C (2017) Robust submap-based probabilistic inconsistency detection for multi-robot mapping. In: 2017 European conference on mobile robots (ECMR), pp 1–6
5. Saeedi S, Paull L, Trentini M, Seto M (2014) Group mapping: a topological approach to map merging for multiple robots. Robot Autom Mag IEEE 21(2):60–72
6. Carpin S (2008) Fast and accurate map merging for multi-robot systems. Auton Robots 25(3):305–316

7. Zhang J, Zhang R, Yue Y, Yang C, Wen M, Wang D (2019) Slat-calib: extrinsic calibration between a sparse 3d lidar and a limited-fov low-resolution thermal camera. In: 2019 IEEE international conference on robotics and biomimetics (ROBIO), pp 648–653. https://doi.org/10.1109/ROBIO49542.2019.8961462
8. Khaleghi B, Khamis A, Karray FO, Razavi SN (2013) Multisensor data fusion: a review of the state-of-the-art. Inf Fusion 14(1):28–44
9. Bonanni TM, Corte BD, Grisetti G (2017) 3-D map merging on pose graphs. IEEE Robot Autom Lett 2(2):1031–1038. https://doi.org/10.1109/LRA.2017.2655139
10. Brand C, Schu MJ, Hirsch H (2015) Submap matching for stereo-vision based indoor/outdoor SLAM. In: IEEE/RSJ international conference on intelligent robots and systems, pp 5670–5677
11. Yue Y, Wang D, Senarathne P, Moratuwage D (2016) A hybrid probabilistic and point set registration approach for fusion of 3D occupancy grid maps. In: 2016 IEEE international conference on systems, man, and cybernetics, pp 1975–1980
12. Hornung A, Wurm KM, Bennewitz M, Stachniss C, Burgard W (2013) OctoMap: an efficient probabilistic 3d mapping framework based on octrees. Auton Robots
13. Zhang J, Siritanawan P, Yue Y, Yang C, Wen M, Wang D (2018) A two-step method for extrinsic calibration between a sparse 3d lidar and a thermal camera. In: 2018 15th international conference on control, automation, robotics and vision (ICARCV), IEEE
14. Redmon J, Farhadi A (2018) Yolov3: An incremental improvement. arXiv preprint arXiv:180402767
15. Yue Y, Senarathne PN, Yang C, Zhang J, Wen M, Wang D (2018) Hierarchical probabilistic fusion framework for matching and merging of 3-d occupancy maps. IEEE Sens J 18(21):8933–8949
16. Zhang J, Singh S (2014) Loam: lidar odometry and mapping in real-time. In: Robotics: science and systems conference
17. Jessup J, Givigi SN, Beaulieu A (2017) Robust and efficient multirobot 3-d mapping merging with octree-based occupancy grids. IEEE Syst J 11(3):1723–1732
18. Rusu RB, Cousins S (2011) 3D is here: point cloud library (PCL). In: IEEE international conference on robotics and automation (ICRA), Shanghai, China

Chapter 7
Collaborative Probabilistic Semantic Mapping Using CNN

Performing collaborative semantic mapping is a critical challenge for cooperative robots to enhance their comprehensive contextual understanding of the surroundings. The chapter bridges the gap between the advances in collaborative geometry mapping that relies on pure geometry information fusion, and single robot semantic mapping that focuses on integrating continuous raw sensor data. In this chapter, a novel hierarchical collaborative probabilistic semantic mapping framework is proposed, where the problem is formulated in a distributed setting. The key novelty of this work is the modelling of the hierarchical semantic map fusion framework and its mathematical derivation of its probability decomposition. At single robot level, the semantic point cloud is obtained by a fusion model combining information from heterogeneous sensors and is used to generate local semantic maps. At collaborative robots level, local maps are shared among robots for global semantic map fusion. Since the voxel correspondence is unknown in collaborative robots level, an Expectation-Maximization approach is proposed to estimate the hidden data association. Then, Bayesian rule is applied to perform semantic and occupancy probability update. The experimental results on the UAV (Unmanned Aerial Vehicle) and UGV (Unmanned Ground Vehicle) platforms show the high quality global semantic map, demonstrating the accuracy and utility in practical missions.

7.1 Introduction

In previous chapters, various challenges in collaborative perception, localization and mapping are carefully addressed. However, these approaches only focus on geometry information, which limit the application of robots in high level tasks. To enhance the perception capability of a group robots, individual robots have to share

Y. Yue and D. Wang, *Collaborative Perception, Localization and Mapping for Autonomous Systems*, Springer Tracts in Autonomous Systems 2,
https://doi.org/10.1007/978-981-15-8860-0_7

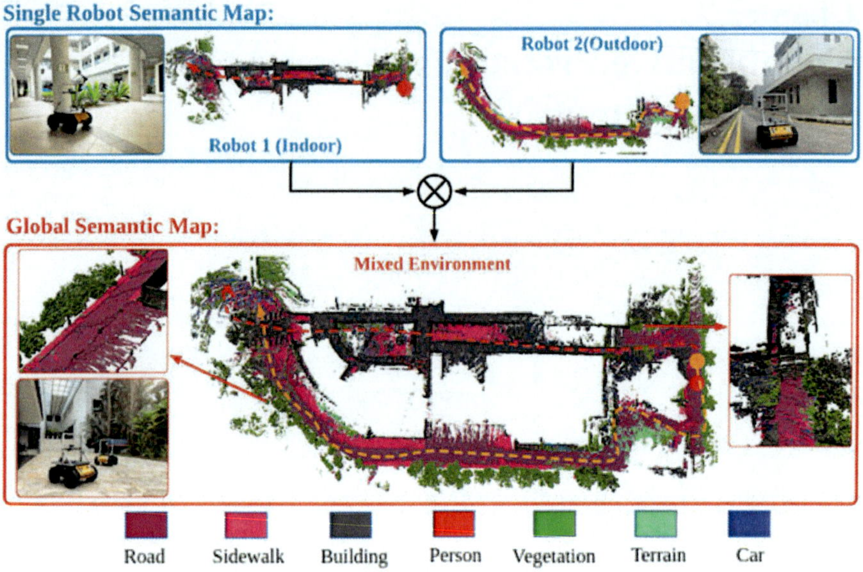

Fig. 7.1 A demonstration of collaborative semantic mapping in mixed indoor-outdoor environment. Red trajectory is robot 1, and orange trajectory is robot 2. Top: local semantic maps generated by two robots. Bottom: collaboratively generated global semantic map

local maps to generate a global representation composed of geometry information and semantic context. Till now, the problem of collaboratively generate a consistent global semantic map using teams of robots remains open. This chapter takes one step forward by focusing on collaborative probabilistic semantic mapping (see Fig. 7.1).

Single robot semantic mapping serves as the basic module for collaborative semantic mapping [1]. In recent years, multimodal sensor-based [2] and Octree-based [3] algorithms have been proposed for single robot semantic mapping. However, there is a fundamental difference between single robot mapping and collaborative mapping because they work on different data levels. Single robot mapping aims to fuse the input continuous raw sensor scans into a local map, while collaborative mapping takes local maps generated by different robots as input. This nature of data heterogeneity makes single-layer structure unable to meet the needs of collaborative mapping. In addition, a single robot does not need to communicate with other robots. However, it is not feasible to transmit raw sensor data between robots with limited bandwidth. Therefore, the first key challenge is how to design a hierarchical collaborative mapping framework that is able to perform collaborative semantic perception efficiently under communication constraints.

Various approaches have been proposed to perform collaborative geometry mapping, such as estimating relative localization [4], accounting for transformation uncertainty [5], updating maps in the global perspective [6]. However, existing approaches in collaborative geometry mapping focus on fusing geometry information such as planes, lines, and points [7]. In the process of geometry information fusion, they

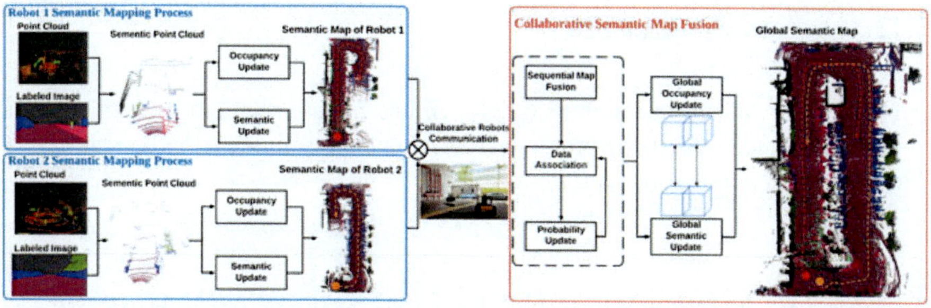

Fig. 7.2 The framework of hierarchical collaborative semantic 3D mapping. Red trajectory is robot 1, and orange trajectory is robot 2

simply stitch the map data or take the average, but did not consider the fusion of semantic information. Collaborative semantic mapping differs from them because the data association between semantic voxels is unknown, which brings another layer of difficulty. In summary, the second key challenge is how to estimate the hidden data association between semantic voxels, update and integrate the probability of correct matching pairs into a consistent global semantic map.

This work is motivated by the fact that comprehensive analysis, modelling and implementation of collaborative semantic mapping methods have not been studied in-depth. The chapter bridges the gap between collaborative geometry mapping that relies on pure geometry information fusion, and single robot semantic mapping that focuses on integrating continuous raw sensor data. The key novelty of this work is the modelling of the hierarchical semantic map fusion framework and its mathematical derivation of its probability decomposition. In the fusion process, an Expectation-Maximization (EM) approach is proposed to estimate the hidden data association. The main contributions are listed as follows:

- A hierarchical collaborative probabilistic semantic mapping framework is proposed and formulated in both single-robot and collaborative robots levels.
- An Expectation-Maximization (EM) algorithm is proposed to estimate the hidden data association between voxels in local maps, where Bayesian rule is applied to perform semantic and occupancy probability update.
- Collaborative system is developed to perform scene understanding and global semantic mapping, demonstrating its accuracy and utility in practical tasks.

The rest of the chapter is organized as follows. Section 7.2 gives an overview of the proposed framework. Section 7.3 explains the theoretical basis for collaborative semantic mapping. Section 7.4 shows the experimental procedures and results. Section 7.5 concludes the chapter with a discussion on future work.

7.2 System Framework

In this section, the system architecture for simultaneous collaborative semantic mapping is presented, where the problem is formulated to provide a theoretical basis for the system on single robot and multi-robot levels.

7.2.1 The Framework of Hierarchical Semantic 3D Mapping

The objective of this chapter is to develop a hierarchical framework for collaborative semantic mapping that enables a group of robots to generate a consistent global semantic map. The overview of the system architecture is depicted in Fig. 7.2, where the framework consists of three modules: *multimodal semantic information fusion, single robot semantic mapping and collaborative semantic map fusion.* Since the proposed multi-robot system operates in a distributed setup, the organization of the chapter follows hierarchical structure from the single robot level to multi-robot level.

For single robot level, the heterogeneous sensors carried by each robot are calibrated and integrated. The robots generate semantic point clouds based on image segmentation and sensor fusion outputs. By applying the Bayesian rule, semantic label probability and occupancy probability are updated to generate single-robot semantic map.

For multi-robot level, each robot communicates with neighboring robots to share local 3D semantic maps. Expectation-Maximization (EM) approach is applied to estimate the hidden data association between voxels in local maps. Then, the local semantic maps are fused into a global consistent semantic map. The main notational symbols are defined in Table 7.1.

7.2.2 Centralized Problem Formulation

Considering a group of robots moving through an unknown environment and attempting to map out the surroundings, the problem can be defined as follows:

Centralized Definition: *Given a group of robots* $\mathcal{R} = \{r\}_{1:R}$, *the objective is to estimate the global semantic map* M_t *given camera observations* $I_{1:t}^{(\mathcal{R})}$, *3D laser observations* $L_{1:t}^{(\mathcal{R})}$ *and robot* trajectories $x_{1:t}^{(\mathcal{R})}$.

$$p(M_t | I_{1:t}^{(\mathcal{R})}, L_{1:t}^{(\mathcal{R})}, x_{1:t}^{(\mathcal{R})}) \tag{7.1}$$

The global semantic map $M = \{M_i\}_1^N$ consists of a set of voxels. Each voxel $M_i = (M_x^i, M_y^i, M_z^i, v_i, o_i)$ is defined as a tuple that includes the position of the extracted voxel center $m_i = (M_x^i, M_y^i, M_z^i)$, the occupancy probability value v_i, the semantic label o_i. The label o_i is assumed to come from a finite set of discrete

Table 7.1 The main notations used throughout the chapter

Symbol	Description
	Single Robot Level:
r	A single robot is denoted as r
$I_t^{(r)}$	Camera observation of robot r at time t
$L_t^{(r)}$	3D laser observation of robot r at time t
$L_{st}^{(r)}$	Semantic labels correspond to 3D laser points
$x_t^{(r)}$	Pose of robot r at time t
$m_t^{(r)}$	Local semantic map generated by robot r at time t
	Multi-Robot Level:
$\mathcal{R}$	Set of all robots $\mathcal{R} = \{r\}_{1:R}$, R is number of robots
$\mathcal{R}_r$	Set of neighboring robots $\mathcal{R}_r = \{r_n\}_{1:R_n}$ in
	communication with r, where the number is R_n
M_r	The global map generated by robot r
T_{r,r_n}	Relative position between robot r and r_n

class labels: $S = \{1, 2, \cdots, k, \cdots\}$. We have $x_t \in SE(3)$ in 3D, $I_t \in \mathbb{R}^2$ in 2D and $L_t \in \mathbb{R}^3$ in 3D.

In the centralized setting, the problem corresponds to calculate the Maximum A Posterior (MAP) estimation problem Eq. (7.1). Assuming the communication is perfect, all the robots can share their latest sensor observations $L_{1:t}^{(\mathcal{R})}$ and $I_{1:t}^{(\mathcal{R})}$ to the control station in real time. However, it is challenging to transfer large size raw sensor data under limited bandwidth.

7.2.3 Distributed Hierarchical Definition

In a constraint environment, each robot r only has access to its own raw sensor observations. In this case, the chapter introduces the single robot layer to perform local semantic mapping, which is an intermediate level that acts as perception and commutation node. In order to improve network efficiency and robustness, only the generated local semantic maps are allocated in the group of robots. The collaborative robots level turns into estimating the global semantic map given the outputs from single robot.

Single Robot Level Definition: *For each robot r, the objective is to estimate its local semantic map $m_t^{(r)}$ given its camera observations $I_{1:t}^{(r)}$, 3D laser observations $L_{1:t}^{(r)}$ and robot trajectory $x_{1:t}^{(r)}$.*

$$p(m_t^{(r)} | I_{1:t}^{(r)}, L_{1:t}^{(r)}, x_{1:t}^{(r)}) \tag{7.2}$$

In single robot level, multimodal semantic information fusion serves as the input for single robot level as well as the whole framework. As defined in Table 7.1, camera observation $I_{1:t}^{(r)}$ and 3D laser observation $L_{1:t}^{(r)}$ servers as the input. Prior to single robot semantic mapping, the multimodal information fusion algorithm is proposed to generate semantic point cloud. The process of multimodal environmental perception is presented in Sect. 7.3.1.

Given the input of semantic point cloud, each robot will estimate its 3D semantic map. For semantic mapping, the occupancy probability and semantic label probability are updated simultaneously. Based on conditional independence, each voxel in the semantic map is updated separately with occupancy update model and semantic update model, obtaining local semantic map $m_t^{(r)}$. The details of single robot semantic mapping are presented in Sect. 7.3.2.

The benefits of introducing the single robot level comes at the following advantages, firstly, the size of the local semantic map is much more compressed compared to the raw sensor data (point cloud and image), which significantly reduces the communication burden; Besides, the factorization also gives the flexibility of choosing various SLAM and semantic segmentation algorithms based on sensor and environment types.

Collaborative Robots Level Definition: *The objective of collaborative semantic mapping is to estimate global semantic map M_r under a fully distributed network, given local maps $m_t^{(r,\mathcal{R}_r)}$ from neighboring robots $\mathcal{R}_r$.*

$$p(M_r|m_t^{(r)}, \Phi_{r_n \in \mathcal{R}_r}(m_t^{(r_n)})) \tag{7.3}$$

At the collaborative robots level, it is assumed that each robot r has estimated its semantic map $m_t^{(r)}$ based on local observations. However, the perception ability and operation area of a single robot are limited. The comprehensive understanding of the environment should be obtained by integrating information from neighboring robots. Hence, robot r receives the local map $m_t^{(r_n)}$ from all the nearby robots $r_n \in \mathcal{R}_r$, where $\mathcal{R} = \{r \cup \mathcal{R}_r\}$ if the communication covers all robots.

Under limited communication, robot r receives neighboring maps in a certain time interval. Hence, the map fusion is performed serially. Given the received local maps, an Expectation-Maximization (EM) approach is proposed to estimate the hidden data association between voxels in local maps, where Bayesian rule is applied to perform semantic and occupancy probability update (see Sect. 7.3.3).

7.3 Collaborative Semantic 3D Mapping

This section presents a distributed hierarchical collaborative semantic mapping framework, which is divided into three subsections, i.e., multimodal semantic information fusion, single robot semantic mapping and collaborative semantic map fusion.

7.3.1 Multimodal Semantic Information Fusion

The goal of multi-modal semantic information fusion is to assign a semantic label for each point in a 3D point cloud. In this work, we utilize the 3D LiDAR and the visual camera as the main sensors for semantic mapping. At time t, to obtain a semantic point cloud, first we generate the semantic image I_{s_t} by passing the raw image I_t through the Deeplab model [8]. Deeplab can output 19 semantic classes, including road, tree, building, and pedestrian etc. Then, we assign the semantic labels from the semantic image to the point cloud by utilizing the calibration parameters and projection equation: first, the 3D point cloud L_t is projected onto the semantic image I_{s_t}, using the projection equation Eq. (7.4). Here, L_t^i is a 3D point in the point cloud (i.e., $L_t^i \in L_t$), T_l^c is the extrinsic parameter between the 3D LiDAR and the camera, K_c is the intrinsic parameter of the camera, l_t^i is the projected point [9]. After projection, the projected point l_t^i is overlapped with a 2D pixel point $p_{s_t}^i$ in the semantic image. Therefore, the semantic label of pixel $p_{s_t}^i$ can be assigned to point l_t^i and L_t^i. Thus, each 3D point L_t^i receives the semantic information. Subsequently, a semantic point cloud L_{s_t} can be obtained (Fig. 7.3).

$$l_t^i = K_c T_l^c L_t^i \tag{7.4}$$

For time 1 to t, the semantic point cloud consists of $L_{1:t}^{(r)}$ and $L_{s\,1:t}^{(r)}$, where $L_{1:t}^{(r)}$ denotes 3D geometry coordinate and $L_{s\,1:t}^{(r)}$ represents the corresponding $19D$ semantic labels, the subscript $1:t$ means time 1 to t, the superscript (r) denotes the robot number r . In the semantic mapping process, the raw images $I_{1:t}^r$ can not be used to generate 3D map directly. Therefore, the semantic point cloud will serve as the

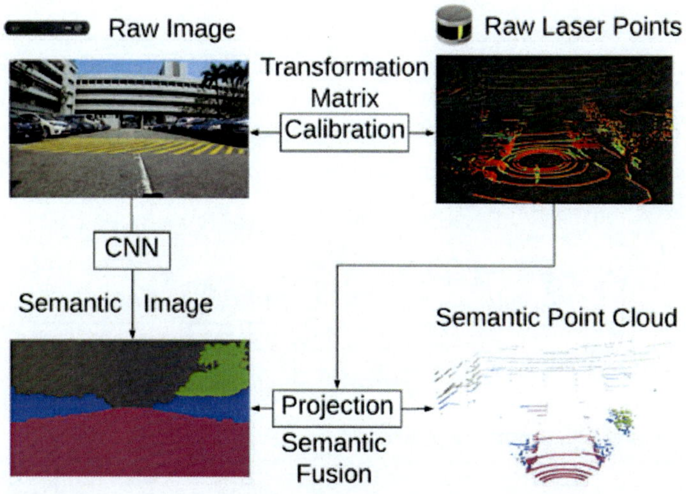

Fig. 7.3 Multimodal semantic information fusion model

input. Then, Eq. (7.2) can be rewritten as:

$$p(m_t^{(r)}|L_{s\,1:t}^{(r)}, L_{1:t}^{(r)}, x_{1:t}^{(r)}) \tag{7.5}$$

7.3.2 Single Robot Semantic Mapping

Single robot localization problem is the basis for semantic mapping, which has been well studied and many efficient approaches have been developed. Any SLAM method can be applied to estimate the odometry $x_{1:t}$, as long as the localization accuracy is guaranteed. As the laser scan is with respect to its local sensor frame, the semantic point cloud is converted into the common coordinate to generate 3D map using the odometry $x_{1:t}$, where $\hat{L}_t = x_t * L_t$ denotes the transformed 3D geometry coordinate, $\hat{L}_{s_t} = x_t * L_{s_t}$ denotes the transformed semantic labels, then we have

$$p(m_t^{(r)}|L_{s\,1:t}^{(r)}, L_{1:t}^{(r)}, x_{1:t}^{(r)}) = p(m_t^{(r)}|\hat{L}_{1:t}^{(r)}, \hat{L}_{s\,1:t}^{(r)}) \tag{7.6}$$

As a result, the probability distribution of map m_r is estimated given observation $\hat{L}_{1:t}^{(r)}$ and $\hat{L}_{s\,1:t}^{(r)}$ as shown in Eq. (7.6). Here, the denotation of $m_t^{(r)}$ is simplified as m_r. For the local semantic map, we define $m_r = \{m_r^i\}_{i=1}^{N_{m_r}}$, where N_{m_r} is the number of voxels in the map. Since the voxel-wise correspondences are usually assumed to be independent, the probability density function of $p(m_t^{(r)}|I_{1:t}^{(r)}, L_{1:t}^{(r)}, x_{1:t}^{(r)})$ can be factorized as:

$$p(m_r|\hat{L}_{s\,1:t}^{(r)}, \hat{L}_{1:t}^{(r)}) = \prod_{i=1}^{N_{m_r}} p(m_r^i|\hat{L}_{s\,1:t}^{(r)}, \hat{L}_{1:t}^{(r)}) \tag{7.7}$$

Then, the probability of a leaf node m_r^i given observation is estimated by updating semantic label probability and occupancy probability simultaneously. Based on conditional independence, each voxel is updated separately with occupancy update model $p(v_r^i|\hat{L}_{1:t}^{(r)})$ and semantic update model $p(o_r^i|\hat{L}_{s\,1:t}^{(r)})$, then Eq. (7.7) is rewritten as Eq. (7.8).

$$\prod_{i=1}^{N_{m_r}} p(v_r^i, o_r^i|\hat{L}_{s\,1:t}^{(r)}, \hat{L}_{1:t}^{(r)}) = \prod_{i=1}^{N_{m_r}} \underbrace{p(v_r^i|\hat{L}_{1:t}^{(r)})}_{\text{Occupancy Update}} \cdot \underbrace{p(o_r^i|\hat{L}_{s\,1:t}^{(r)})}_{\text{Semantic Update}} \tag{7.8}$$

7.3.2.1 Voxel Occupancy Update

The occupancy probability is recursively updated given the incoming 3D point cloud $\hat{L}_{1:t}^{(r)}$. Based on Bayesian rule, the occupancy update model is expanded and updated in Eq. (7.9). The updating of Eq. (7.8) depends on the current observation $p(v_r^i|\hat{L}_t^{(r)})$, the previous estimation $p(v_r^i|\hat{L}_{1:t-1}^{(r)})$ and the prior probability $p(v_r^i)$. $p(v_r^i)$ denotes the initial occupancy probability of each voxel and is set as 0.5.

$$p(v_r^i|\hat{L}_{1:t}^{(r)}) = \left[1 + \frac{1 - p(v_r^i|\hat{L}_t^{(r)})}{p(v_r^i|\hat{L}_t^{(r)})} \frac{1 - p(v_r^i|\hat{L}_{1:t-1}^{(r)})}{p(v_r^i|\hat{L}_{1:t-1}^{(r)})} \frac{p(v_r^i)}{1 - p(v_r^i)}\right]^{-1} \tag{7.9}$$

7.3.2.2 Semantic Label Update

The semantic probability is recursively updated given the incoming set of 19 semantic labels $\hat{L}_{s\,1:t}^{(r)} = \{\hat{L}_{s\,1:t}^{(r)}(k)\}_{k=1}^{19}$. Then, $p(o_r^i|\hat{L}_{s\,1:t}^{(r)})$ denotes the probability of updated label class (k=1:19). Based on Bayesian rule, the semantic update model is expanded and updated in Eq. (7.10). The updating of Eq. (7.10) depends on the current observation $p(o_r^i|\hat{L}_{s_t}^{(r)})$, the previous estimation $p(o_r^i|\hat{L}_{s\,1:t-1}^{(r)})$ and the prior probability $p(o_r^i)$. $p(o_r^i)$ denotes the initial semantic probability of each voxel and is set as $\frac{1}{19}$.

$$p(o_r^i|\hat{L}_{s\,1:t}^{(r)}) = \left[1 + \frac{1 - p(o_r^i|\hat{L}_{s_t}^{(r)})}{p(o_r^i|\hat{L}_{s_t}^{(r)})} \frac{1 - p(o_r^i|\hat{L}_{s\,1:t-1}^{(r)})}{p(o_r^i|\hat{L}_{s\,1:t-1}^{(r)})} \frac{p(o_r^i)}{1 - p(o_r^i)}\right]^{-1} \tag{7.10}$$

After fusion, the class that corresponds to maximum probability is assigned as the label of the voxel. The process of single robot semantic map update is shown in Fig. 7.4a, b.

7.3.3 Collaborative Semantic Map Fusion

7.3.3.1 Sequential Semantic Map Fusion

Under limited bandwidth condition, maps are generated and transmitted sequentially in a certain time interval. This results in robot r receiving its neighboring robots' maps in some permutation π. Hence, the map fusion is performed serially, where a certain threshold is satisfied to trigger a pair-wise map sharing and fusion. Then, Eq. (7.3) can be factorized into Eq. (7.11). Initially, the relative transformation matrix T_{r,r_n} between m_r and m_{r_n} is unknown. To estimate T_{r,r_n}, the map matching algorithm proposed in [10] is used. Then, the neighboring robot map m_{r_n} is transformed to coordinate frame of robot r by transformation function $\Phi(T_{r,r_n}, m_{r_n})$, which is simplified as $\Phi(m_{r_n})$.

$$p(M_r|m_r, \Phi_{r_n \in \mathcal{R}_r}(m_{r_n})) = \prod_{r_n=1}^{N_r} p(M_r^{\pi(1:r_n)}|M_{r_n}^{\pi(1:r_n-1)}, \Phi(m_{\pi(r_n)})) \tag{7.11}$$

Note that $M_r^{\pi(1,r_n)}$ is the partial global map of robot r generated by fusing the maps of robots $\pi(1)$ to $\pi(r_n) \in \mathcal{R}_r$ with m_r after the reception of map $m_{\pi(r_n)}$ from neighboring robot $\pi(r_n) \in M_r^{\mathcal{R}_r}$. The final global fused map $M_r^{\mathcal{R}_r}$ of robot r can be retrieved after the end of this serial process, hence $M_r^{\mathcal{R}_r} = M_r^{\pi(1,Nr)}$.

7.3.3.2 Global Map Occupancy and Label Update

The key difference between local and global semantic map update comes at two hands. First, the input of single robot mapping is raw sensor observation (Fig. 7.4a), while the input of global semantic mapping is local semantic maps (Fig. 7.4c). Since voxel correspondence is unknown, we need to establish the data association relationship before fusion. Second, the same object can be observed in different perspectives by different robots, the voxels representing the same object can have different semantic class. To show the difference, an example of is presented in Fig. 7.5. In the left image, the grey color denotes geometry 3D map. As can be seen, the geometry map fusion is only based on closet neighborhood distance search, which ignores the object attributes. In contrast, the right image shows an example of semantic map fusion, where blue denotes car and green represents grass. In this case, semantic data association should reject the wrong correspondences and accepts correct pairs by incorporating semantic information.

Hence it is vital to come up with a strategy to consider the dissimilarities and fuse them into the global semantic map (Fig. 7.4d). The following two subsections will detail the hidden data association estimation and global information fusion.

For simplification, we denote the latest incoming neighboring robot map as m_{r_n} and current robot map as m_r. Therefore, the probability distribution in Eq. (7.11) is simplified as $p(M_r|m_r, m_{r_n})$. Since the voxel-wise correspondences are usually assumed to be independent, the probability density function of $p(M_r|m_r, m_{r_n})$ can be factorized into Eq. (7.12). When m_r^i and $m_{r_n}^j$ are matched voxels, they are fused

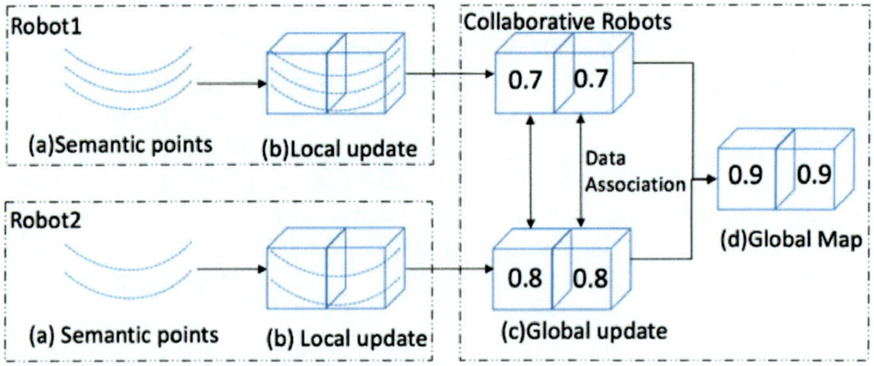

Fig. 7.4 An example of local and global semantic map update process. The value represents semantic probability, indicating the semantic update process

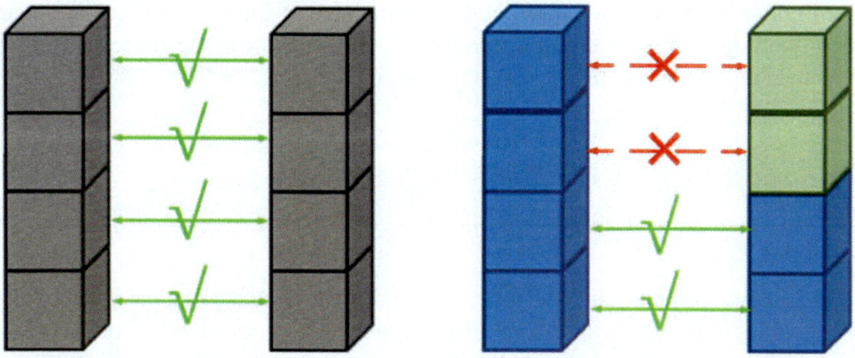

(a) Geometry map fusion (b) Semantic map fusion

Fig. 7.5 An example of geometry map fusion process in Fig. 7.5a, it only considers the geometry information and simply stitch the map. While in Fig. 7.5b, semantic information is considered to establish correct data associations. The red cross rejects the wrong data association

in pairwise to generated fused voxel M_r^k.

$$p(M_r|m_r, m_{r_n}) = \prod_{i=1} p(M_r^k|m_r^i, m_{r_n}^j) \tag{7.12}$$

To estimate the hidden data association between m_r^i and voxel $m_{r_n}^j$. A binary variable $d_{i,j}$ is introduced to represent the data association between corresponding voxels m_r^i and $m_{r_n}^j$, where $d_{i,j} = 1$ if m_r^i corresponds to $m_{r_n}^j$ and equals 0 otherwise. Based on total probability theory, Eq. (7.12) is rewrote in the form of Eq. (7.13).

$$p(M_r|m_r, m_{r_n}) = \prod_{i=1}^{n} \sum_{j} p(M_r^k, d_{i,j}|m_r^i, m_{r_n}^j) \tag{7.13}$$

To solve Eq. (7.13), Expectation Maximization (EM) is applied, which is an algorithm well-suited for models containing latent variables. There are two steps for EM: the E-step efficiently estimates the hidden variables by evaluating the expectation, while M-step updates the global map probability given the corresponding voxel pairs (see Eq. (7.14)).

$$p(M_r^k, d_{i,j}|m_r^i, m_{r_n}^j) = \underbrace{p(d_{i,j}|m_r^i, m_{r_n}^j)}_{\text{E step:data association}} \cdot \underbrace{p(M_r^k|d_{i,j}, m_r^i, m_{r_n}^j)}_{\text{M step:probability update}} \tag{7.14}$$

E-Step: The E-step establishes the correspondence by calculating the minimum relative distance. Here, we define a 5D descriptor as $\{m_i^x, m_i^y, m_i^z, v_{m_i}, o_{m_i}\}$, which includes the center coordinate of voxel m_i^x, m_i^y, m_i^z $(3D)$, occupancy probability v_{m_i} $(1D)$ and semantic labeling o_{m_i} $(1D)$. The corresponding voxel is calculated by finding the nearest neighborhood as formulated in Eq. (7.15). Once the nearest neighborhood within a certain threshold is found, the data association is established between m_r^i and $m_{r_n}^j$ with $d_{i,j} = 1$. Otherwise, we let $d_{i,j} = 0$.

$$p(d_{i,j}|m_r^i, m_{r_n}^j) = d_e(m_r^i, m_{r_n}^j) + d_v(v_{m_r^i}, v_{m_{r_n}^j}) + o_v(o_{m_r^i}, o_{m_{r_n}^j})$$
$$= ||m_r^i - m_{r_n}^j||^2 + ||v_{m_r^i} - v_{m_s^j}||^2 + ||o_{m_r^i} - o_{m_s^j}||^2 \tag{7.15}$$

M-Step: Given the data association $d_{i,j}$ and corresponding voxel pairs $m_r^i, m_{r_n}^j$, the next step is to fuse corresponding voxels into the global voxel M_r^k. Equation (7.16) is the M-step of updating the probability, which is defined in Eq. (7.14). Since each voxel contains occupancy probability and semantic label probability, Eq. (7.16) is rewritten into Eq. (7.17). Considering the occupancy update and semantic update are updated independently, Eq. (7.18) is factorized as the global occupancy update model and global semantic update model.

$$p(M_r^k|d_{i,j}, m_r^i, m_{r_n}^j) \tag{7.16}$$
$$= p(v(M_r^k), o(M_r^k)|d_{i,j}, v(m_r^i), v(m_{r_n}^j), o(m_r^i), o(m_{r_n}^j)) \tag{7.17}$$
$$= \underbrace{p(v(M_r^k)|d_{i,j}, v(m_r^i), v(m_{r_n}^j))}_{\text{Global Occupancy Update}} \cdot \underbrace{p(o(M_r^k)|d_{i,j}, o(m_r^i), o(m_{r_n}^j))}_{\text{Global Semantic Update}} \tag{7.18}$$

In single robot occupancy updating process Eq. (7.9), the input is a set of raw semantic points. However, the input in collaborative robots level is the probability value of local semantic map. Each voxel can be regraded as a Gaussian distribution, which is located at the center of the voxel and identical in each direction. In this case, the fusion at collaborative robots level is the integration of two Gaussian distributions. The initial semantic probability of each global map voxel $p(v(M_r^k))$ is 0.5. Then, the global occupancy update is formulated in Eq. (7.19).

$$p(v(M_r^k)|d_{i,j}, v(m_r^i), v(m_{r_n}^j)) = \left[1 + \frac{1 - p(v(m_{r_n}^j))}{p(v(m_{r_n}^j))} \frac{1 - p(v(m_r^i))}{p(v(m_r^i))} \frac{p(v(M_r^k))}{1 - p(M_r^k)}\right]^{-1} \tag{7.19}$$

For global semantic label probability update, the input is 19 class probabilities of each voxel in single robot $o(m_r^i)$, $o(m_{r_n}^j)$). The initial semantic probability of each global map voxel $p(o(M_r^k))$ is $\frac{1}{19}$ (see Eq. (7.20)).

$$p(o(M_r^k)|d_{i,j}, o(m_r^i), o(m_{r_n}^j)) = \left[1 + \frac{1 - p(o(m_{r_n}^j))}{p(o(m_{r_n}^j))} \frac{1 - p(o(m_r^i))}{p(o(m_r^i))} \frac{p(o(M_r^k))}{1 - p(o(M_r^k))}\right]^{-1} \tag{7.20}$$

After fusion, the probability value for all 19 labels are updated. The semantic probability of the most likely class $p(o(M_r^k), max)$ for each voxel M_r^k is computed as follows:

$$p(o(M_r^k), max) = \arg\max[p(o(M_r^k), 1), \ldots, p(o(M_r^k), 19)] \tag{7.21}$$

As a result, the class that corresponds to maximum probability is assigned to the class label of the voxel.

7.4 Experimental Results

The performance of the proposed collaborative semantic mapping framework is validated through extensive real-world experiments. Both qualitative and quantitative results are presented. The former is obtained by visualizing the fused map in various experimental settings. Then, the latter is expressed by showing average entropy, update process and size of data.

7.4.1 Evaluation Overview

7.4.1.1 Experimental Platform

Experiments are conducted by operating the UGV platforms in real environments. All algorithms are executed on the ROS platform. Two UGV platforms (Husky Clearpath) are equipped with the Intel Core i7-6700HQ CPU @ 2.60GHz CPU and the Nvidia GeForce RTX 2060 @ 6GB RAM GPU. The UGVs are also equipped with a 3D Velodyne LiDAR and a visual camera, where the sensors have already been calibrated. The UAV platform is designed by ourselves and is equipped with the ZED stereo camera with Intel NUC 6i7KYK @ 2.60GHz CPU. The communication between robots is established by long range Wi-Fi with limited bandwidth.

For the software packages, LOAM [14] is applied for UGV pose estimation. ORB [15] is applied for UAV pose estimation. The Cityscapes dataset [16] is used to trained semantic segmentation model. HPFF [10] is implemented to estimate relative transformation between robots, and Octomap [17] is utilized for basic 3D geometry mapping with a resolution of 0.2m.

7.4.1.2 Experiment Setup

Experiments are conducted by operating UGV or UAV in three different scenarios in Nanyang Technological University (Fig. 7.6). All algorithms are executed on the

(a) Open Carpark (b) Mixed Environment (c) UAV-UGV Mapping

Fig. 7.6 The experiment setting up: Open Carpark, Mixed Environment, UAV-UGV Mapping respectively

ROS platform for multimodal semantic information fusion, single robot semantic mapping and collaborative semantic map fusion.

- **Open Carpark**: two Husky robots equipped with Velodyne 3D LiDAR and visual camera in an open carpark.
- **Mixed Environment**: two Husky robots equipped with Velodyne 3D LiDAR and visual camera in an indoor-outdoor mixed environment.
- **UAV-UGV Semantic Mapping**: one Husky robot equipped with Velodyne 3D LiDAR and visual camera, one UAV equipped with ZED stereo camera.

7.4.1.3 Comparison Baseline

Most of the existing approaches either focus on single robot semantic mapping or collaborative geometry mapping. A direct qualitative comparison of our method with other approaches is conducted and summarized in Table 7.2. The existing work can achieve part of the functions, but this chapter is the only work that addresses the collaborative semantic mapping problem and realizes all functions. As no available work has addressed collaborative semantic mapping problem, we first demonstrate extensive qualitative results of how the global map is updated and how the local maps

Table 7.2 Comparison of core functionality with existing approaches

Algorithms	Category	Multimodal Perception	Geometry Mapping	Semantic Mapping	Collaborative Fusion	Distributed Communication
Yang et al. [11]	Single Robot	×	✓	✓	×	×
Berrio et al. [3]	Single Robot	✓	✓	✓	×	×
Saeedi et al. [12]	Multiple Robots	×	✓	×	✓	×
Jessup et al. [13]	Multiple Robots	×	✓	×	✓	×
This work	Multiple Robots	✓	✓	✓	✓	✓

are fused. Then, we present quantitative results on average entropy, update process, size of data and computation time.

7.4.2 *Open Carpark*

To evaluate the multimodal semantic fusion algorithm for single robot semantic mapping, we first test our system in the Open Carpark. After processing the input data, semantic 3D reconstruction results can be obtained, which is presented in Fig. 7.7. The figure shows top view of the map and three close-up views for each time step. As shown in Fig. 7.7, the algorithm successfully integrates and updates the semantic information into 3D semantic map. More specifically, our approach can reconstruct the 3D map and recognize the classes of objects on the road with high accuracy.

The collaborative semantic mapping results is shown in Fig. 7.8. Two robots started their mapping from nearby places and explored the environment by traversing two different paths. Figure 7.8 shows the fusion of maps at start, middle and final of the mission. The left two columns shows two local semantic maps. The right side is the fused global semantic map. For the entire process of collaborative mapping, it is shown that our system performs well in both single-robot and multi-robot levels. For single-robot level, the semantic map can be completely built by utilizing the raw sensor data perceived by each robot. For collaborative robots, the algorithm achieves the fusion of maps from two robots. The algorithm combines the map information from two maps to generate a consistent global map.

7.4.3 *Mixed Indoor Outdoor*

The mixed environment presented in Fig. 7.9 is a building in the university, where the indoor environment is the lobby, and the outdoor environment is the outside road. Robot **1** travelled through the lobby in the building, and robot **2** moved along the road outside the building. The two robots first started nearby in the initial position, it can be seen that the two maps are successfully merged into a global map. Then, two robots encountered at the end position, and generated a global semantic map. Thanks to the probability update formulation, the overlapped area combines the advantages between the two maps, resulting in a detailed global map.

7.4.4 *UAV-UGV Mapping*

This part presents the result of UAV-UGV collaborative semantic mapping. It differs from the previous two experiments since heterogeneous robots and sensors are

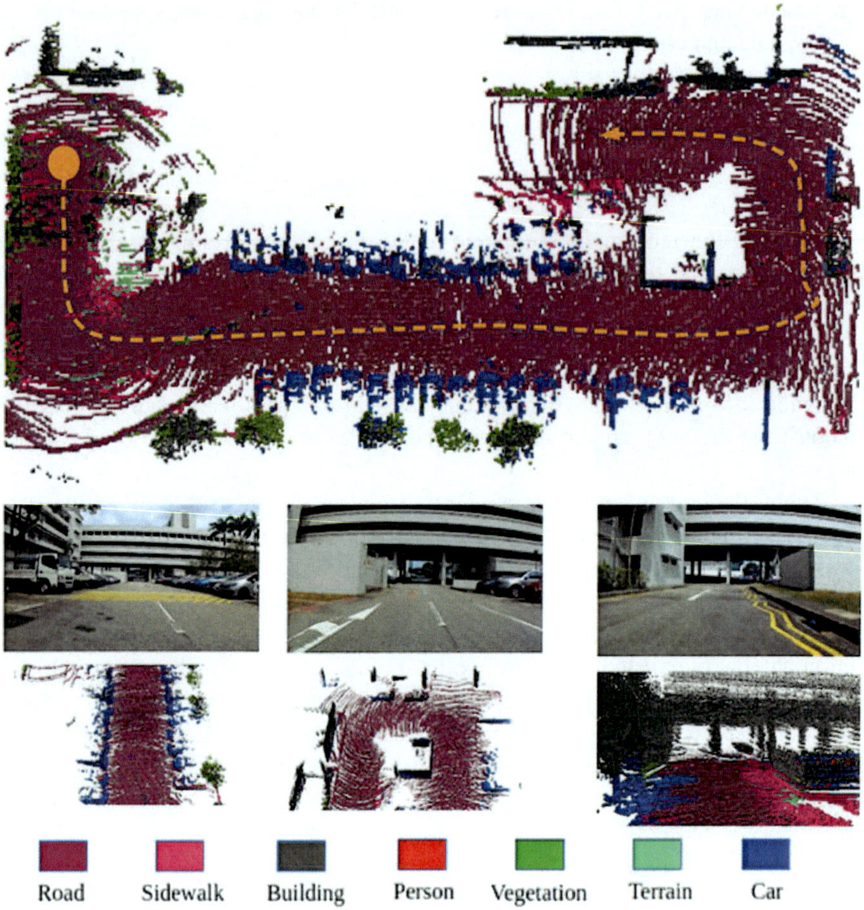

Fig. 7.7 Robot 2 semantic map in Open Carpark. Only the seven classes that appear in the scene are marked

applied. In this environment, two robots started from a nearby place. The UGV could not cross the staircase, while the UAV directly flew over the staircase giving its high mobility (see Fig. 7.6c). The UAV is equipped with ZED stereo camera, which performs semantic segmentation and generates point cloud. Figure 7.10 demonstrates the semantic maps generated by UAV in two scenarios.

As presented in Fig. 7.11, UAV and UGV generated local semantic maps at the end of the mission. Due to the specification of sensors, UAV generated a dense semantic map and successfully mapped the staircase. UGV captured most of the environmental information. Due to the limited mobility, the UGV still failed to climb the staircase and map out the area as shown in Fig. 7.11. By combing the local semantic maps generated by UAV and UGV, a more detailed global semantic map is reconstructed. It

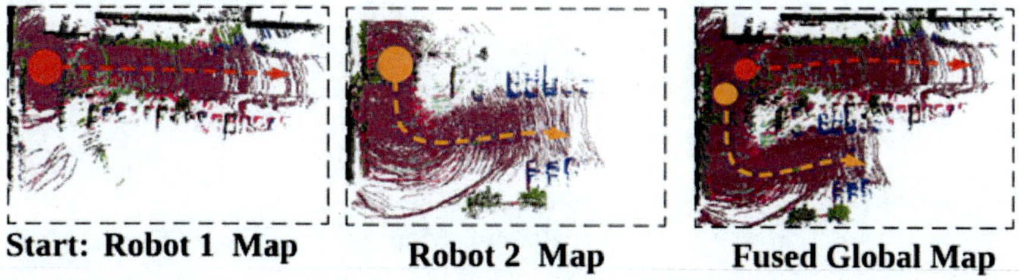

(a) t=0s, the initial position of the two robots. Two sparse local maps are successfully fused into a global map.

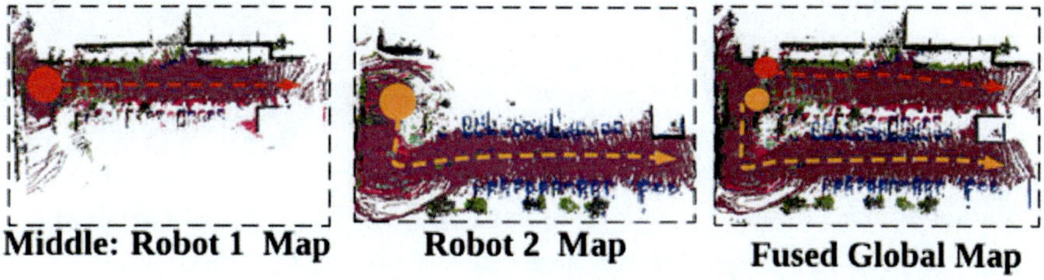

(b) Half of the mission. The local maps become denser, and global map are fused.

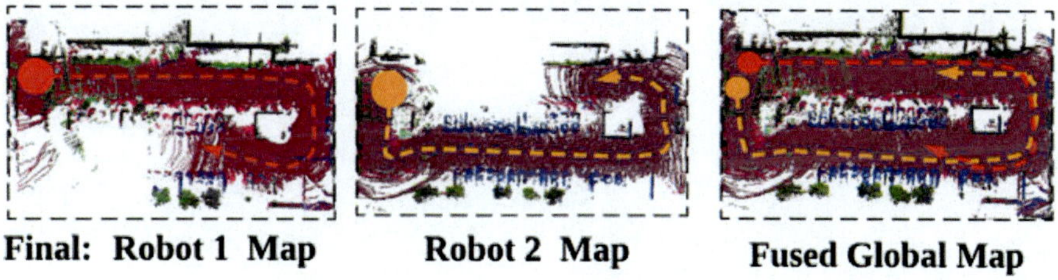

(c) End of the mission. The global map of the full environment is presented.

Fig. 7.8 Collaborative semantic mapping at the start, middle and final of the task in the open carpark

shows the flexibility of the proposed algorithm in different robot platforms equipped with heterogeneous sensors.

7.4.5 Quantitative Analysis

First, the average entropy is calculated to show the improved map uncertainty. We also examine the fusion accuracy in the global mapping process. Finally, the data sizes of the three experiments are summarized and compared.

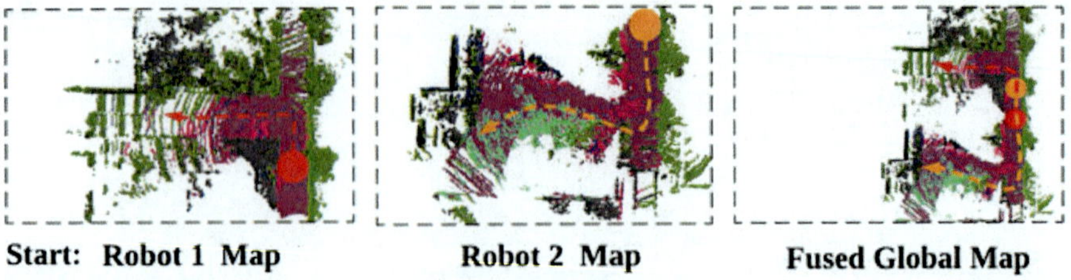

(a) t=0s, the initial position of the two robots. Two sparse local maps are successfully fused into a global map.

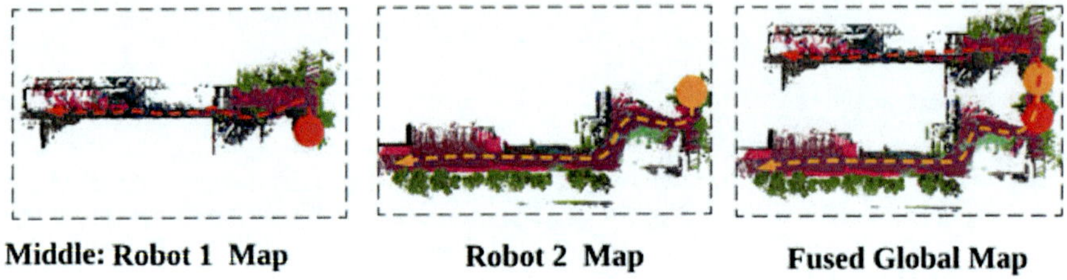

(b) Half of the mission. The local maps become denser, and global map are fused.

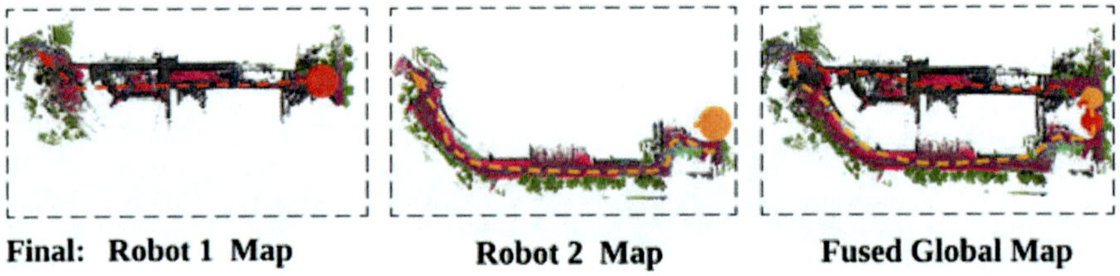

(c) End of the mission. The global map of the full environment is presented.

Fig. 7.9 Collaborative semantic mapping at the start and final of the task in the mixed environment

7.4.5.1 Average Uncertainty

Table 7.5 summarizes the average entropy of the maps before and after collaborative mapping. In entropy theory, the higher the probability of the map, the lower the corresponding entropy. Therefore, we use average entropy to measure the uncertainty of the map. In Table 7.5, the value presents the average entropy in different experiments. For occupancy probability update, the entropy has been reduced after performing the proposed algorithm. For label probability update, the entropy has been also reduced. Then, the average entropy of overall probability update process decreased up to 10% in all experiments. Experiments show that the proposed map fusion strategy can effectively combine the probability information of two maps, thereby reducing the uncertainty of occupancy probability and label probability.

Fig. 7.10 The semantic map generated by UAV in two scenarios

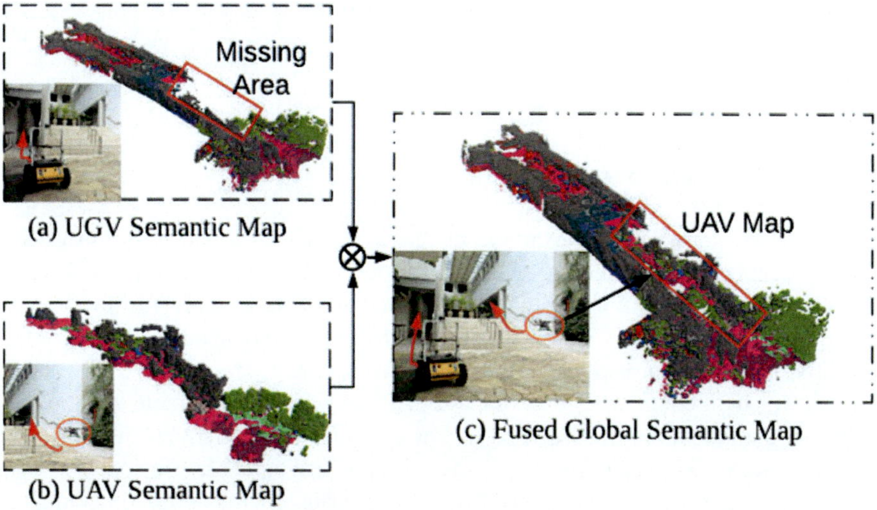

Fig. 7.11 The result of UAV-UGV semantic mapping. **a**, **b** Single robot SLAM is perform to generate the local semantic map. **c** The fused global semantic map

Table 7.3 Global map label fusion (Correct False Positive (FP) Label)

	Robot1 (FP)		Robot2 (TP)		Global (TP)	
No	Label	Probability	Label	Probability	Label	Probability
1	Road	0.638	Car	0.973	Car	**0.894**
2	Road	0.711	Car	0.985	Car	**0.969**
3	Road	0.834	Car	0.992	Car	**0.977**

Table 7.4 Global map label fusion (Enhance True Positive (TP) Label)

	Robot1 (TP)		Robot2 (TP)		Global (TP)	
No	Label	Probability	Label	Probability	Label	Probability
1	Road	0.695	Road	0.773	Road	**0.784**
2	Road	0.810	Road	0.855	Road	**0.886**
3	Road	0.906	Road	0.924	Road	**0.943**

Table 7.5 Average entropy of the maps in all experiments. (The best performance is denoted in bold.)

Experiments	Occupancy Probability			Label Probability			Overall Probability		
	Map 1	Map 2	Global Map	Map 1	Map 2	Global Map	Map 1	Map 2	Global Map
Open Carpark	0.4030	0.3929	**0.3712**	0.3689	0.3653	**0.3429**	0.3439	0.3267	**0.3078**
WKW Lobby	0.3088	0.3285	**0.2849**	0.3239	0.3327	**0.2937**	0.2833	0.2930	**0.2715**
UAV-UGV	0.3291	0.3476	**0.3161**	0.3349	0.3527	**0.3229**	0.3098	0.3157	**0.2895**

7.4.5.2 Updating Process

Table 7.3 demonstrates the example of fusing mutual information from two local maps. In this scenario, the ground truth for the voxels from robot **1** and **2** should be Car. It is worth noting that the voxel (ground truth label is Car) from robot **1** is assigned with the class of Road with a low probability of 0.638, however this is a false positive (FP). The corresponding voxel in the robot **2** has a true positive (TP) class of Car with a high confidence level of 0.973. As can be seen from Table 7.3, the voxel label in the global map is assigned to Car after fusion. This shows the superiority of the proposed algorithm for correcting false positive label after fusion. Then, Fig. 7.12 shows more detailed semantic probability update process of this voxel in global map, specifying the label and its corresponding probability. We can observe the trend of increasing of the label Car (blue) and the decreasing in the remaining labels.

Table 7.4 shows the example of fusion process can enhance semantic probability of true positive label. The corresponding voxels from two local maps indicate the true class label of Road. The fusion integrates the probability information and increases the semantic probability in the global map.

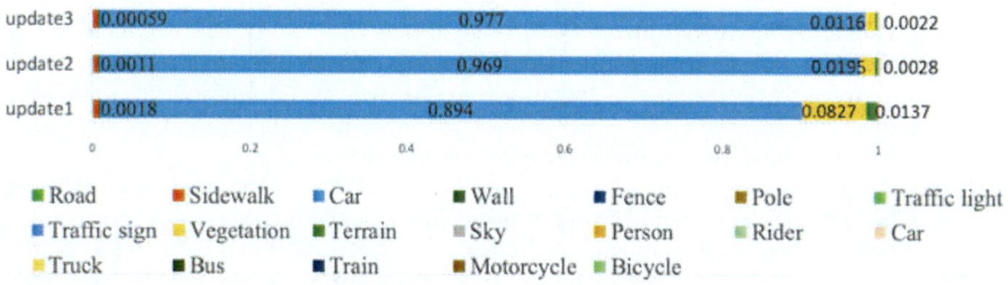

Fig. 7.12 Semantic probability update process of a voxel in global map

7.5 Conclusion

This chapter established a hierarchical collaborative probabilistic semantic mapping framework. We have designed a new framework to provide theoretical formulas and system implementations for collaborative semantic mapping. In the single robot level, the semantic point cloud was obtained by a fusion model combining information from heterogeneous sensors and used to generate local semantic maps. To achieve collaborative semantic mapping, this chapter provided theoretical basis and algorithm implementation for the global 3D semantic mapping. The results have shown the algorithm was able to establish the correct data association between voxels. More importantly, the fusion process was able to correct the false label and enhance true label. The overall experimental results presented high quality global semantic maps, demonstrated the accuracy and utility of the framework. In summary, the proposed collaboration system provides a new perspective for sensing and adapting to real environments, which compensates for the limitations of individual robot perception and mapping.

In the future, relative localization and place recognition of collaborative robots can be performed on the basis of global semantic map, which will significantly improve the autonomy of the robots.

References

1. Kostavelis I, Gasteratos A (2015) Semantic mapping for mobile robotics tasks: a survey. Robot Auton Syst 66:86–103
2. Yue Y, Zhao C, Li R, Zhang J, Wen M, Yuanzhe W, Wang D (2020) A hierarchical framework for collaborative probabilistic semantic mapping. In: 2020 IEEE international conference on robotics and automation (ICRA)
3. Berrio JS, Zhou W, Ward J, Worrall S, Nebot E (2018) Octree map based on sparse point cloud and heuristic probability distribution for labeled images. In: 2018 IEEE/RSJ international conference on intelligent robots and systems (IROS), IEEE, pp 3174–3181
4. Chen H, Sun D, Yang J, Chen J (2009) Localization for multirobot formations in indoor environment. IEEE/ASME Trans Mechatron 15(4):561–574

5. Yue Y, Yang C, Wang Y, Senarathne PCN, Zhang J, Wen M, Wang D (2019) A multilevel fusion system for multirobot 3-d mapping using heterogeneous sensors. IEEE Syst J 14(1):1341–1352
6. Yue Y, Wang D, Senarathne P, Moratuwage D (2016) A hybrid probabilistic and point set registration approach for fusion of 3D occupancy grid maps. In: 2016 IEEE international conference on systems, man, and cybernetics, pp 1975–1980
7. Yue Y, Yang C, Zhang J, Wen M, Wu Z, Zhang H, Wang D (2020) Day and night collaborative dynamic mapping in unstructured environment based on multimodal sensors. In: 2020 IEEE international conference on robotics and automation (ICRA)
8. Badrinarayanan V, Kendall A, Cipolla R (2017) Segnet: a deep convolutional encoder-decoder architecture for image segmentation. IEEE Trans Pattern Anal Mach Intell 39(12):2481–2495
9. Zhang J, Siritanawan P, Yue Y, Yang C, Wen M, Wang D (2018) A two-step method for extrinsic calibration between a sparse 3d lidar and a thermal camera. In: 2018 15th international conference on control, automation, robotics and vision (ICARCV), IEEE
10. Yue Y, Senarathne PN, Yang C, Zhang J, Wen M, Wang D (2018) Hierarchical probabilistic fusion framework for matching and merging of 3-d occupancy maps. IEEE Sens J 18(21):8933–8949
11. Yang S, Huang Y, Scherer S (2017) Semantic 3d occupancy mapping through efficient high order CRFs. In: 2017 IEEE/RSJ international conference on intelligent robots and systems (IROS), IEEE, pp 590–597
12. Saeedi S, Paull L, Trentini M, Seto M (2014) Group mapping: a topological approach to map merging for multiple robots. Robot Autom Mag IEEE 21(2):60–72
13. Jessup J, Givigi SN, Beaulieu A (2015) Robust and efficient multirobot 3-D mapping merging with octree-based occupancy grids. IEEE Syst J PP(99):1–10
14. Zhang J, Singh S (2014) Loam: lidar odometry and mapping in real-time. In: Robotics: science and systems conference
15. Mur-Artal R, Montiel JMM, TardÃşs JD (2015) Orb-slam: a versatile and accurate monocular slam system. IEEE Trans Robot 31(5):1147–1163
16. Cordts M (2016) The cityscapes dataset for semantic urban scene understanding. In: 2016 IEEE conference on computer vision and pattern recognition (CVPR), pp 3213–3223
17. Hornung A, Wurm KM, Bennewitz M, Stachniss C, Burgard W (2013) OctoMap: an efficient probabilistic 3d mapping framework based on octrees. Auton Robots

Chapter 8
Conclusions

8.1 Summary

This book aims to develop systematic framework for collaborative perception, localization and mapping for multiple robots. The book provides comprehensive theoretical formulations, algorithmic development, and experimental verification via simulation and realtime experiments. The book first proposes the probabilistic point set registration algorithm to match and merge 3D maps. In structured environment, a novel hierarchical probabilistic map fusion framework is proposed to combine the high level geometric features and low level voxels, which improves the convergence basin and robustness. For robots equipped with heterogeneous sensors in unstructured environments, a general probabilistic framework for multi-robot 3D mapping is validated in various scenarios, which is independent of sensor types and SLAM algorithms. In a challenging all-weather and dynamic environment, day and night collaborative dynamic mapping is proposed based on multimodal sensors. Finally, with the rapid development of convolutional neural networks, this book promotes collaborative perception to a higher semantic level and proposes a collaborative semantic mapping algorithm. The main contributions and conclusions of this book are summarized as follows:

- Firstly, the book addressed the probabilistic point set registration approach for fusion of 3D occupancy grid maps (Chap. 3). The proposed occupancy iterative closet point (OICP) algorithm combines positional coordinate in conjunction with occupancy probability to increase the matching accuracy, where the local descriptor is augmented from 3D space to 4D space. As the proposed OICP doesn't guarantee global minima, the chapter has applied environment measurement model to evaluate the quality of the transformation. The final map fusion is then achieved with a relative entropy filter which integrates the measurements of both maps and decreases the uncertainty of the global map. Experiments on simulated and real world environments were conducted and the evaluations confirmed that the pro-

Y. Yue and D. Wang, *Collaborative Perception, Localization and Mapping for Autonomous Systems*, Springer Tracts in Autonomous Systems 2,
https://doi.org/10.1007/978-981-15-8860-0_8

posed OCIP method is able to produce more consistent global 3D occupancy grid maps compared to existing methods in the literatures.

- Secondly, the book proposed hierarchical probabilistic fusion framework for matching and merging of 3D occupancy maps (Chap. 4). This chapter studied the hierarchical probabilistic fusion framework for 3D occupancy maps, which is factorized into three modules: uncertainty modeling, map matching, and map merging. The local uncertainty of the map is modeled by combing the individual voxel and structural edge covariances. Based on the uncertainty modeling, two-level Probabilistic Map Matching (PMM) algorithm is applied to calculate the relative transformation accurately. In the PMM, it combines both high-level structural edge and low-level occupancy voxel features. Experiments on simulated and real-world environments are conducted, and the evaluations confirm the utility of the proposed approach in generating more accurate relative transformation with a faster rate and a larger convergence basin.
- Thirdly, the book presented general probabilistic framework for collaborative 3D mapping using heterogeneous sensors (Chap. 5). This chapter has provided a general probabilistic framework for collaborative 3D mapping, which is independent of sensor types and SLAM algorithms. In the book, the Expectation-maximization (EM) based map matching algorithm is successfully applied to register 3D Lidar map and Kinect map, where the numbers of voxels to be associated is automatically decided based on the map characteristics. To distribute the maps among robots, the uncertainties of relative localization estimation and the local map error have been propagated and integrated. The results verify the accuracy, efficiency and generality for collaborative mapping missions. The proposed mapping framework can be utilized for complexed environment mapping for real time multiple mobile robots applications.
- Fourthly, the book established an approach for collaborative dynamic mapping in day and night unstructured environments (Chap. 6). Operating in day and night environment is challenging for multiple robots, especially in the presence of dynamic objects. This chapter designed a novel framework to provide theoretical formulas and system implementations for collaborative dynamic mapping in all weather environment. The method was evaluated during day and night in unstructured environment, which exhibited improved precision, efficiency and robustness compared to existing methods.
- Lastly, the book presented a hierarchical collaborative probabilistic semantic mapping framework (Chap. 7). We have designed a new framework to provide theoretical formulas and system implementations for collaborative semantic mapping. To achieve collaborative semantic mapping, this chapter provided theoretical basis and algorithm implementation for the global 3D semantic mapping. The overall experimental results presented high quality global semantic maps, demonstrated the accuracy and utility of the framework. The proposed collaboration system provides a new perspective for sensing and adapting to real environments, which compensates for the limitations of individual robot perception and mapping.

8.2 Open Challenges

This book has developed a systematic framework for collaborative perception, localization and mapping. Although collaborative system has been investigated in depth in this book, there are still several interesting research problems that are worthy to be studied.

The problem of matching accuracy and convergence is a key issue for collaborative system. The proposed map matching algorithm belongs to the scope of the iterative optimization algorithm, which tends to be trapped into a local minimum. Thus, more improvements on finding the initial inputs should be explored. For example, GPS information can also be fused to support the initial estimation if available. Since an erroneous initial input would further undermine the accurate matching, hypothesis testing strategy to reject the wrong estimation should be developed. In addition, the availability of a consistent global map requires periodic fusion of partial maps from multiple robots. Hence making a decision on when to fuse the maps warrants further research and would also be explored in the future.

The collaboration between heterogeneous robots is another problem worth studying. For hybrid teams robots (UGV and UAV), the large difference in viewpoint and robot kinematics brings the challenge of fusion different sensor modalities generated by heterogeneous robots. The difficulty comes at finding the corresponding regions that allow an accurate relative localization between the heterogeneous robots. Low level features like lines and points are often view-point dependent and subject to failure in repetitive environments. With the fast developing in deep learning, it is promising to conduct Convolutional Neural Networks-based approach to assign correspondence at semantic levels, which will increase the robustness and efficiency.

Current 3D geometry mapping approaches only contain geometry information, which limit the application of robots in high level tasks. To enhance the perception capability of a group robots, individual robots have to share local maps to generate a global representation composed of geometry information and semantic context. Semantic map is useful in robots formation for the traversability analysis (roads or bushes) in unstructured environments. For a group of service robots, it can also be used to assign intelligent tasks, for example, robot 1 delivers food to the room, and 2 patrols outside the house. For autonomous vehicles, sharing semantic maps among them will avoid accidents caused by occlusion, and update the global semantic map of the city more efficiently.

In the near future, the autonomous systems are expected to interact each other, share their maps, perform relative localization and accomplish a complex mission by assigning tasks autonomously. The above vision requires efforts in the fields of robotics, artificial intelligence, control and computer science. The authors will be a part of these exploration directions, and we hope to see the whole community join us.